Operations Readiness & Assurance (OR&A®)

A Guide for Practitioners

David C. Powell

Copyright ©2012 by David Powell. All rights reserved.

Published by: Operations Readiness & Assurance (OR&A®) Ltd.

No part of this publication may be reproduced, stored in a retrieval system, or transmitted in any form or by any means, electronic, mechanical, photocopying, recording, scanning, or otherwise, without the prior written permission of the Author.

The information in this book is intended for use as a learning and information resource for an individual purchaser of the book. Use of the techniques, methodologies and strategies presented in this book by companies or commercial organisations is not permitted without prior written permission from the Author and OR&A® Ltd.

The acronym OR&A® and the OR&A logo are registered trademarks. Reproduction of these by companies or commercial organisations is not permitted without prior written permission from the Author and OR&A® Ltd.

Limit of Liability / Disclaimer of Warranty:

Whilst the author and publisher have used their best efforts in preparing this book, they make no representations or warranties with respect to the accuracy or completeness of the contents of this book and specifically disclaim any implied warranties of merchantability or fitness for a particular purpose.

No warranty may be created or extended by sales representatives or agents within any written sales materials.

The advice and strategies contained in this book may not be suitable for all situations. Professional advice should therefore be sought before use of any such advice or strategy, as appropriate.

Neither the publisher nor author shall be liable for any loss of profit, cost or any other commercial damages, including, but not limited to special, incidental, consequential or other damages.

Powell, David C.
Operations Readiness & Assurance (OR&A), A Guide for Practitioners

ISBN 978-1-4716-0802-5

2012

Table of Contents

1. Management Structures — 11
   - 1.1 Where It All Started — 11
   - 1.2 The Industrial Revolution — 12
     - 1.2.1 Frederick Winslow Taylor — 12
     - 1.2.2 Henri Fayol — 14
     - 1.2.3 Karol Adamiecki — 16
     - 1.2.4 Henry Gantt — 17
     - 1.2.5 Walter A. Shewhart — 18
     - 1.2.6 Dr W. Edwards Deming — 19
     - 1.2.7 Ronald A. Fisher — 21
     - 1.2.8 Edward W. Merrow — 22
   - 1.3 In The Last 50 Years — 23
   - 1.4 Assurance — 26
     - 1.4.1 Definition of Assurance — 26
     - 1.4.2 Inherent Assurance — 27
     - 1.4.3 Real Assurance — 27
     - 1.4.4 Achieving Real Assurance — 28

2. Mega-Projects or Mega-Failures — 31
   - 2.1 Identifying the Challenges — 31
   - 2.2 Asset Operations Team — 32
   - 2.3 Project Timeline — 33
   - 2.4 Front End Loading (FEL) — 34
   - 2.5 Business Case for Using OR&A — 35
   - 2.6 CAPEX vs. OPEX — 37
   - 2.7 Introducing OR&A — 38
   - 2.8 The New Car Analogy — 39
     - 2.8.1 Similarities in the Processes — 39
     - 2.8.2 Operations Requirements — 39

3. Introduction to OR&A — 45
   - 3.1 Definition of OR&A — 45
   - 3.2 OR&A Pre-Requisites — 46
     - 3.2.1 Management Commitment — 46
     - 3.2.2 Management Structure — 46
     - 3.2.3 OR&A Team Competencies/Experience — 47

4. Developing an OR&A Capability — 49
   - 4.1 Management of the OR&A Process — 49
     - 4.1.1 Policy — 49
     - 4.1.2 Organisation — 50
     - 4.1.3 Planning & Implementation — 50
     - 4.1.4 Performance Measurement — 50
     - 4.1.5 Performance Review — 50
     - 4.1.6 Audit — 50

	4.2	The OR&A Information Structure	51
		4.2.1 Corporate Information Structure	51
		4.2.2 OR&A Information Structure	53
5	OR&A System Structure		55
	5.1	The Work Breakdown Structure (WBS)	55
		5.1.1 Generic OR&A Groups	55
		5.1.2 OR&A Subjects	56
		5.1.3 Activities	57
		5.1.4	57
		5.1.5 Tasks	58
		5.1.6 Deliverables	59
	5.2	Populating the OR&A Structure	60
		5.2.1 Delivery Maps	60
		5.2.2 Detail Sheets	60
		5.2.3 Supporting Information (Library)	62
6	OR&A System Content		63
	6.1	OR&A Management (1.0)	63
		6.1.1 OR&A Management (1.01)	63
		6.1.2 Cost Estimates (OPEX) & Management (1.02)	67
		6.1.3 Quality Management (1.03)	73
		6.1.4 Licences, Permits & Permissions Management (1.04)	77
		6.1.5 Flawless Project Delivery (1.05)	79
	6.2	Operations Management (2.0)	82
		6.2.1 Operations Management & Planning (2.01)	82
		6.2.2 Production Operations (2.02)	86
		6.2.3 Production Chemistry & Flow Assurance (2.03)	90
		6.2.4 Wells (2.04)	105
		6.2.5 Infield & Export Pipelines & Manifolds (2.05)	108
		6.2.6 Sub-Sea Systems (2.06)	111
		6.2.7 Maintenance & Integrity Management (2.07)	113
		6.2.8 Process Automation & Control Systems (2.08)	119
		6.2.9 Infrastructure, Logistics & Geomatics (2.09)	125
		6.2.10 Materials Procurement & Management (2.10)	128
		6.2.11 IM, IT & Communications Systems (2.11)	131
		6.2.12 Contracted Services (2.12)	135
	6.3	Operations Organisation & Competency (3.0)	138
		6.3.1 Operations Organisation (3.01)	138
		6.3.2 Operations Training & Competency (3.02)	142
	6.4	HSS&E Management (4.0)	144
		6.4.1 Health, Safety & Environmental (HSE) Mgmt. (4.01)	144
		6.4.2 Security Management & Community Relations (4.0)	150
		6.4.3 Risk Management (4.03)	154
	6.5	Commissioning & Start-Up Management (5.0)	160
		6.5.1 Commissioning & Start-Up (CSU) (5.01)	160
		6.5.2 Asset Handover & Acceptance (5.02)	164

7	Implementing OR&A on a Project	167
	7.1 Dissemination of Information	167
	7.2 Setting-Up the OR&A Application	168
	7.2.1 Introduction	168
	7.2.2 Definitions	168
8	Measuring and Monitoring OR&A	171
	8.1 Recording OR&A Progress/Status	171
	8.1.1 Monitoring Progress	172
	8.2 OR&A System Outputs	173
	8.2.1 Preparing an OR&A Plan	173
	8.2.2 OR&A Spreadsheet	174
	8.2.3 The OR&A Dashboard	174
	8.2.4 Tailored Reports	175
	8.2.5 Providing Assurance	175
9	Lessons Learned & Best Practices	179
	9.1 Lessons Learned	179
	9.1.1 Definition of a Lesson Learned	179
	9.1.2 Identification of Lessons Learned	180
	9.1.3 Potential Lessons	181
	9.1.4 Classification	181
	9.1.5 Collection/Retention/Storage	182
	9.1.6 Retrieval/Application	183
	9.1.7 Promotion of Lessons Learned	183
	9.2 Best Practices	184
	9.2.1 Identification	184
	9.2.2 Retrieval/Application	184
	9.2.3 Promotion of Best Practices	184
	9.3 Skeleton Documents	185
	9.3.1 What is a Skeleton Document	185
	9.3.2 Creation	185
	9.3.3 Retrieval/Application	185
	9.4 Knowledge Management	186
	9.4.1 Explicit Knowledge	186
	9.4.2 Tacit Knowledge	186
	9.4.3 Implicit Knowledge	187
	9.4.4 Knowledge Management in the OR&A System	188
10	Further Information / Assistance	189

Table of Figures

Figure 1 -	The 14 Principles of Management	15
Figure 2 -	Lewis' Circular Logic	18
Figure 3 -	Project Phases & Gates	33
Figure 4 -	Influence vs. Cost in Project Phases	34
Figure 5 -	Design Intent vs. Typical Performance	35
Figure 6 -	Life Cycle Costs	37
Figure 7 -	Typical OR&A Competency Matrix	47
Figure 8 -	ISO 9000 Management System	49
Figure 9 -	Corporate Hierarchy	51
Figure 10 -	OR&A Hierarchy	53
Figure 11 -	Decision Gates & Reviews	57
Figure 12 -	Generic OR&A Delivery Map	60
Figure 13 -	Cost Estimate Maturity 'Funnel'	67
Figure 14 -	Flawless Start-Up	79
Figure 15 -	Lessons Learned & Best Practices	180

Preface

This book describes the Operations Readiness & Assurance (OR&A) process as experienced and developed by David Powell over more than 12 years of involvement in an OR&A capacity on a number of Oil & Gas E&P mega-projects (projects with a total value in excess of US$ 100 million).

The book frequently refers to the products and services of Operations Readiness & Assurance (OR&A) Ltd. developed by David and his son Jonathan, that have been created specifically to satisfy the requirements of implementing OR&A on such projects.

You can find out more about OR&A Ltd. including those products and services, by visiting the website at:

http://www.operationsreadinessandassurance.com

or

http://www.or-and-a.com

Key Objective Statement

An Operations Readiness & Assurance (OR&A) team must always strive to make it clear to everyone with whom they work that they are not there to oversee or judge the work of other disciplines.

The purpose of OR&A is only to provide assurance that everything they need (i.e. the Operations Requirements) to operate the asset will be ready at the point of handover of the ownership of the asset to them.

This requires the OR&A Team to clearly identify those needs at the earliest stage of the project and to follow the development of those needs (providing advice and guidance wherever it is needed) to ensure those Operations Requirements are satisfied.

1 Management Structures

1.1 Where It All Started

The Ancient Egyptians have laid claim to being the first civilised race to want to standardise the way they did things, so around 3,000 BC, they came up with the first measurement standard.

The Cubit (or Mahe), was a unit of measurement based on the length of a human forearm. It was then subdivided into 7 palm widths each four digits wide, making 28 increments in total.

The obvious variations in measurements made using this system, due to the varying physical dimensions of subsequent users, meant that any measurement could only be approximate; not a good quality when you are building a pyramid.

The typical variation in the length of the Cubit was between 523mm and 529mm so provide a single, accurate value, the Royal Egyptian Cubit was carved onto granite slate and wooden copies (or standards) were then prepared.

The standards were rigorously controlled and frequently checked following a schedule based on the lunar calendar. Compliance was also rigorously implemented and failure to comply was punishable by death (a clear example of a standard with a supporting control mechanism).

The evidence supporting the existence of the Royal Egyptian Cubit comes from documents describing the construction of the Step Pyramid at Djoser in the Lower Nile Valley of Egypt during the Third (Old Kingdom) Dynasty in Egypt which lasted from 2,686 B.C. and 2,181 B.C.

The first evidence of controlled business processes was demonstrated by the Shang Dynasty between the 16th and 11th centuries B.C. where the physical tasks involved in the production of a number of handicrafts was divided into specific disciplines and the products were passed along from person to person as each subsequent task was performed as required.

Even the Greek philosopher Aristotle recognised the value of maintaining standards when he said, *'One swallow does not make a spring ... excellence is a habit, not an event'*.

Moving on to Mediaeval Europe, the Guilds of Master Craftsmen set standards and provided training for apprentices to maintain standards and transfer knowledge. In Venice, from around 1320 A.D. a system of mass producing warships was developed which used standardised (interchangeable) components and Gutenberg's invention of moveable type in the 15th Century revolutionised the production of printing books in quantity whilst maintaining quality.

Manufacturing weapons in component form and then assembling them in a production line was first used by Frenchman Honoré Blanc in 1778, (though Eli Whitney, an American arms manufacturer, also claimed to be the first to have done this in 1803).

1.2 The Industrial Revolution

Engineering made great leaps forward at the start of the Industrial Revolution in the early 1800s when steam power and large machinery provided the power and the capability was developed to manufacture components in large quantities, requiring appropriate standards, processes and controls to be developed.

1.2.1 Frederick Winslow Taylor

F. W. Taylor (1856-1915) was an American mechanical engineer who sought to improve industrial efficiency. He is regarded as the originator of scientific management as a discipline and the first person on record to have applied the systematic observation and study of work as a way to achieve that improvement.

He was effectively one of the first management consultants and wrote his methods in a book entitled 'The Principles of Scientific Management' which was first published in 1911. He was often called the Isaac Newton (or perhaps the Archimedes) of the science of work, though he only laid the foundations.

Taylor had some quite extreme views (not unusual in the Victorian era) about the abilities and intellect of the 'working man' and it is no surprise that he had many detractors, some of whom referred disparagingly to his ideas as Taylor's Principles or Taylorism.

Taylor's scientific management consisted of four principles:

- Base work methods on a scientific study of the tasks required.
- Scientifically select, train, and develop each employee.
- Provide detailed instruction (and supervision) for discrete tasks.
- Divide work between managers and workers, so that the managers plan and manage the work and the workers perform the tasks.

Taylor thought that by analysing work, the 'One Best Way' to do it could be identified. He is accredited with inventing the stopwatch time study, which when later combined with Frank Gilbreth's motion study methods became the 'Time and Motion Study', decomposing a job into its component parts and accurately measuring the time to complete each component.

Taylor's ideas were based on three fundamental methods:

- Find the best practice (benchmarking),
- Decompose the work into its constituent tasks (business process),
- Get rid of things that don't work (continuous improvement).

The principles of scientific management proposed by Taylor continue to be valid and are still used in Project Management today.

Project work is all based on the activities of specific disciplines, carried out by experienced, competent personnel following detailed codes, standards and methods and following an Integrated Project Plan developed from a Work Breakdown Structure (showing the work to be carried out and allowing managers to manage).

The ideas Taylor has about the best way of doing work is reflected in our current thinking of identifying Best Practices and Lessons Learned.

Henri Louis Le Chatelier (1850-1936), a French engineer and chemist, translated Taylor's work and introduced scientific management throughout government owned plants during World War I.

This translation of Taylor's work also influenced Henri Fayol (1841-1925), a French theorist and mining engineer who wrote 'Administration Industrielle et Générale' published in 1916, which emphasized organizational structure in management.

In his 1916 book 'Administration Industrielle et Generale', the French Engineering Management Theorist, Henri Fayol wrote that Taylor approached work methods from the 'bottom up' starting with units of activity (tasks), rather than his own perspective of 'top down'.

Taylor devised new methods for making tasks more efficient and applied what he learnt to the whole hierarchy (Lessons Learned?).

Henri Fayol (1841-1925) also comments that Taylor's methods require too many people to identify the ways to improve efficiency and results in a loss of the focus inherent in single line reporting and management techniques.

It is not difficult to realise that without the benefit of modern computer based techniques and applications, this approach would be very resource intensive, resource hungry and difficult to visualise.

Fayol's theories, based on observation and experience, recommend a consistent set of principles that all organizations must apply in order to run properly.

1.2.2 Henri Fayol

Henri Fayol (1841-1925) was one of the most influential contributors to modern concepts of management, having proposed that there are five primary functions of management, (Planning, Organizing, Commanding, Coordinating, and Controlling), though in more recent texts on the subject this has been revised to four functions, (Planning, Organizing, Leading, and Controlling).

Controlling is described in the sense that a manager must receive feedback on a process in order to make necessary adjustments. Fayol's work has stood the test of time and has been shown to be relevant and appropriate to contemporary management.

Based largely on his experience, Fayol also promoted what he called the '14 Principles of Management' which are also key components of most modern personnel and management systems, as described in the following table:

Henri Fayol	OR&A
1. Specialization of labour	Competence and experience
2. Authority	Seniority, roles and responsibilities
3. Discipline	Rigorous process set out in corporate documents
4. Unity of command	Single line reporting structure
5. Unity of direction	A robust system clearly identifying what needs to be done and when
6. Subordination of Individual Interests	Professionalism, enthusiasm
7. Remuneration	Good terms, conditions and prospects
8. Centralization	Clear command structure, roles and responsibilities
9. Chain of Superiors	Single line reporting structure
10. Order	A robust system
11. Equity	Good terms and conditions, opportunity to develop and progress career
12. Personnel Tenure	
13. Initiative	A robust system clearly identifying what needs to be done and when (OR&A Plan)
14. Esprit de corps	Communication, morale, job satisfaction

Figure 1 - The 14 Principles of Management

1.2.3 Karol Adamiecki

The early principles of management were developed by Karol Adamiecki (1866-1933), who graduated in engineering from the University Of St. Petersburg, Russia in 1891 and moved to live in Poland, becoming a professor of the Warsaw Polytechnic in 1922.

Adamiecki published his first papers in scientific management in 1898, before Frederick Winslow Taylor had popularized the subject.

In 1896 Adamiecki invented a novel means of displaying interdependent processes so as to enhance the visibility of production schedules in the steel industry. The theory, published in 1903 in a number of articles in the Polish magazine Przegląd Techniczny (Technical Review) caused a stir in Russian technical circles.

Unfortunately, his work was published in Russian and Polish, languages not widely known at that time in the English speaking world and therefore he did not receive the credit due to him.

In 1925, Adamiecki became the founder of the Institute of Scientific Organization (Instytut Naukowej Organizacji) in Warsaw, Poland and subsequently served as the vice president of the European Association of Scientific Management (Europejskie Stowarzyszenie Naukowego Zarządzania).

By the time Adamiecki did publish an article describing his diagram in 1931, which he called a harmonograf, a similar concept had been popularized in the West by Henry Gantt (who had published articles on it in 1910 and 1915).

Adamiecki is also the author of 'The Law of Harmony in Management', which presents the concept that management harmony should comprise three parts:

- harmony of choice (tools should be mutually compatible)
- harmony of doing (the importance of schedules and timetables)
- harmony of spirit (the importance of teamwork)

Again, although these ideas were revolutionary at the time, they make good sense to us now.

1.2.4 Henry Gantt

Henry Gantt (1861-1919) an engineering graduate of the Stevens Institute of Technology in New Jersey, USA, worked with F. W. Taylor from 1893 until 1893 applying scientific management principles to their work in the steel industry.

In 1903, during his career as a management consultant, Henry Gantt came up with his idea for the Gantt chart, similar to the invention by Adamiecki, which he further developed and popularised between 1910 and 1915.

The Gantt chart is still accepted as an important management tool and provides a graphic schedule for the planning and controlling of work, and for recording progress in the stages of completing a project.

So, we have the principles of Scientific Management which follow Taylor's four principles and effectively describe the Work Breakdown Structure and we have the Gantt chart which introduces the concept of a timeline.

The approach to Operations Readiness & Assurance described in this book builds on concepts, including the work of Karol Adamiecki in that we have adapted his idea of novel means of displaying the interdependent processes so as to enhance their visibility, with the significant benefit of 21st century computer power.

1.2.5 Walter A. Shewhart

In 1924, W. A. Shewhart (1891-1967) created the first process quality control charts by applying statistical theory and methods during research at the Bell Telephone Laboratories and this became an essential tool for predicting and controlling the quality of industrial processes and established the scientific basis for quality control.

Shewhart met Deming 1927, and because of their mutual interests, Shewhart became Deming's mentor and this association is subsequently reflected his approach to applied research via statistical theory and methods. Shewhart's 1931 book, The Economic Control of Quality, had great influence on Deming's ideas and writings.

Clarence Irving Lewis, the Edward Pierce Professor of Philosophy at Harvard at the time and author of the book 'Mind and The World Order: Outline of a Theory of Knowledge' was also an influence on both Shewhart and Deming.

According to Lewis, 'a good logic must be circular' and this is possibly one reason why Shewhart used a circle to illustrate quality logic.

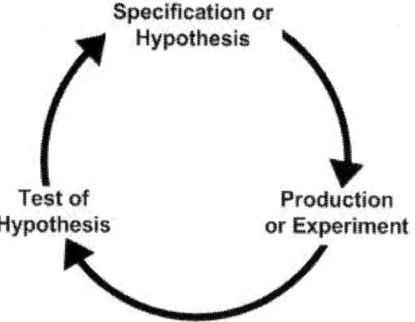

Figure 2 -   Lewis' Circular Logic

Shewhart wrote that the ideal model of continuous quality improvement is best represented as 'a circle that describes a dynamic scientific process via three steps that run parallel to the scientific method' as illustrated below:

This logic provides the basis for what is known as the Shewhart Cycle (& Deming Wheel) for continuous quality improvement which eventually became Plan, Do, Study, Act (or Plan, Do Check, Act for Deming).

1.2.6 Dr W. Edwards Deming

W. E. Deming (1900-1993) is remarkable in that he had worked as a consultant statistician reaching the age of 80 years old before he became generally recognised in the United States of America, as can be determined only by the availability of his personal and professional papers in the American Library of Congress.

These papers demonstrate the influence that Shewhart and others had on Deming and the high esteem in which Shewhart was held by Deming. They give an account of Deming's views on the Plan-Do Study-Act or PDSA cycle, named the Shewhart Cycle, and the Plan-Do-Check-Act or PDCA cycle (which Deming in his written material in the archive apparently disowns).

Dr Deming emerged from relative obscurity in the USA onto the world quality scene after the NBC documentary "If Japan Can, Why Can't We?" was shown on television.

It is clear from his books that Deming viewed management in his country with some concern. He advocated the 'transformation' of American management with the statement that since there is a stable system; quality improvement is the responsibility of management.

His makes it quite clear that the failures lie in bad management. Echoing Taylor, Deming concludes that "Best efforts and hard work, not guided by knowledge, only dig deeper the pit that we are in".

In his writings, Deming also postulates on the aim of leadership, stating: "The aim of leadership should be to improve the performance of man and machine ... Specifically, a leader must learn by calculation, (wherever meaningful figures are at hand), or by judgement otherwise, who, if any, of his people are outside the system ... and hence are in need of either individual help or deserve recognition in some form".

He goes on to stress the responsibility of 'leaders' to 'improve the system' and later he continues "...responsibility is to accomplish greater and greater consistency of performance within the system, so that apparent differences between people continually diminish."

The importance of training in contributing to the fulfilment of these responsibilities is stressed in the following paragraphs.

Deming also writes about a 'system of profound knowledge', which, he suggests "individual components of the system, instead of being competitive, will for optimization reinforce each other for the accomplishment of the aim of the system" and "provides a new map of theory by which to understand and optimize the organisations that we work in".

Careful studies of the writings of Taylor & Deming indicate that there are many similarities between their views and approaches, not least that:

- Both allocate significant responsibility to management for the achievement of organisational objectives,
- Both advocate teamwork,
- Both advocate 'pride in workmanship',
- Both stress the importance of theory and knowledge,
- Both indicate that significant improvement requires outside influences.

The management of change was a further common factor, Taylor warning that the transition to 'scientific management' would take several years and Deming warning against the 'hope for instant pudding'.

It is interesting that the messages that Taylor was conveying at the beginning of the 20th century are so similar to those that Deming was conveying at the end of the century.

Toyota used Deming's ideas in the 1950s and 1960s to transform their company from a sub-par company to one that eventually surpassed American companies like Ford and GM.

Deming's ideas also form the backbone of most of the new management techniques that were introduced at the turn of the millennium.

However, he also observed that in many cases, despite its failure to deliver sustainable prosperity in much of the developed world, the 'initiative and incentive' approach to management continues to prevail.

1.2.7 Ronald A. Fisher

Ronald A. Fisher (1890-1962) was also influential, at least initially, in Deming's statistical thinking. Fisher, like Shewhart and Deming, originally studied mathematics and physics.

Fisher was a prolific scholar who revolutionized statistical theory and methods. In estimation theory, he introduced the concepts of unbiasness, efficiency, consistency, and sufficiency; he first coined the term likelihood, and developed the theory of maximum likelihood estimation.

1.2.8 Edward W. Merrow

Edward W. Merrow is the founder and CEO of Independent Project Analysis, Inc. (IPA), and a recognized expert on the development and execution of large and complex megaprojects and has published several excellent books on the subject.

His broad knowledge of capital projects is sought out by company executives and professionals worldwide. His expertise in megaprojects is built on decades of research on the unique challenges of these complex investments.

With continued megaproject activity across the oil, minerals, and chemical industries in all parts of the world, Merrow's IPA studies are motivated by the industry's need to develop and execute these projects more successfully.

His latest publication on the subject, Industrial Megaprojects - Concepts, Strategies, and Practices for Success (John Wiley & Sons, April 2011) examines how such improvements could be made and has a number of principles common to Project Management and OR&A in particular.

In 2007, Edward W. Merrow received the U.S. Construction Industry Institute's highest honour, the Carroll H. Dunn Award of Excellence, and the 2001 AIChE Engineering and Construction Contracting (ECC) Division Award for Outstanding Contributions to the Industry.

1.3 In The Last 50 Years

The development of some important areas of management thinking and processes in the 20th Century began with the work of Frederick Winslow Taylor and the century ended (almost) with the work of William Edwards Deming.

The common features of their approach were that both were seeking to improve productivity, and both saw it as a management responsibility to make it happen.

Both of these influential thinkers then, had an approach to organisations which placed heavy burdens on the management of the organisation and presented the opportunity to assess whether management was fulfilling their role. 'Systems Thinking' was at the core of the beliefs and approach of both men.

The essence of modern quality management thinking is that the organisation should seek to improve every aspect of its activity. That such improvement is an ongoing and endless task, and that it is the responsibility of all people in the organisation to contribute to this goal.

Some of the elements of the quest for improvement include teamwork, leadership, training, constancy of purpose, sustainability, knowledge of variability, etc. Some of the approaches to this include ISO 9000 series of standards, business excellence models and frameworks and other comprehensive organisation wide analytical processes. The Australian

Taylor also recognised that it was difficult to manage what one could not understand, and his approach to data collection and use was based on the idea that one way of gaining better understanding was to measure and quantify.

In the case of variability, the underlying ideas of the 'one best way' were that if a 'best way' could be found, then it would be advantageous for all of those doing the task to do it that way.

In several publications from 1997 & 1998, which revisited Taylor's written work, gave the first indication that 'standardisation' and 'written instructions' were an important part of the management of variation. Nowadays, we might refer to 'procedures', 'work instructions' and the 'Management of Change' in the case of the ISO 9000 set of standards.

Taylor also provides evidence of recognition of the employer's responsibility to ensure that workers were employed in 'safe' working conditions in order to ensure sustainability of the worker's contribution, where he say that "...in no case is the workman to be called upon to work at a pace which would be injurious to his health ...".

The task is always so regulated that the man who is well suited to his job will thrive while working at this rate during a long term of years and grow happier and more prosperous, instead of being overworked." The emphasis on recruitment of the correct man, appropriate training and development and monitoring of progress are all elements that would be recognised today as good practice.

Taylor also recognised the importance of stakeholders. "At first glance we see only two parties to the transaction, the workmen and their employers. We overlook the third great party, the whole people, - the consumers, who buy the product of the first two and who ultimately pay both the wages of the workmen and the profits of the employers." The introduction to the book begins with a statement about natural resources and the importance of not squandering the non-renewable resources.

Taylor likens the inefficient way of working to the needless and senseless squandering of those natural resources. It is, I think, unknown whether he related inefficient work practices with the use of much greater amounts of natural resources than would be required in a more efficient workplace.

In the last 50 years, many companies and organisations have worked to develop an OR&A system, mostly for military or mission capability, not least the US Department of Energy who published a paper on Operational Readiness in 1987.

The work of Edward W. Merrow brings together many of the methodologies, best practices and lessons learned identified over the years and neatly connects them to the world of mega-projects.

The OR&A system described in this book embraces many of these ideas, but the key to this OR&A process is the way in which it employs the various technologies that were simply not available until recently (since the advent of the desktop PC in the mid-1980s).

Combining the visual display capabilities, the power of the computer database and software systems and the clear project management logic presented by Edward W. Merrow and his predecessors has allowed us to create a very powerful OR&A system.

However, none of these very capable people managed to address the issue of assurance, especially the provision of assurance demanded by the latest OR&A systems.

1.4 Assurance

1.4.1 Definition of Assurance

The component of corporate governance by which a management team provides accurate and current information to the stakeholders about the efficiency and effectiveness of its policies and operations, and the status of its compliance with the statutory obligations is generically termed Assurance.

Within Oil & Gas Projects, this is often expressed as a statement of the degree of confidence that a given specific status has been achieved by the Project Team.

In the context of Operations Readiness & Assurance, it refers to the output from the OR&A System that provides a level of assurance to the stakeholders stating whether that asset (and its supporting organisation) is at a given level of Readiness to Operate (or will reach it by the required date).

Such a statement of Assurance must be based on two specific values:

- The target value of what constitutes 'Readiness for Operations', such a definition statement or other clear defined value,
- A mechanism to measure the value of that 'Readiness for Operations' in a meaningful manner.

Without these components, any statement of assurance is reduced to an implied status of assurance reported by the Project Team and the level of that assurance perceived by the stakeholders from those reports. In reality, this assurance is only **Inherent Assurance** based mostly on trust and experience.

However, applying these components in a structured and rigorous way to determine a statement of assurance by logical measurement of actual status provides **Real Assurance** based on measured values.

Such assurance has the added benefit of identifying the source of any inadequacies allowing prompt and effective remedial action.

Manual measurement, recording and analysis of the values required to provide that Real Assurance by traditional reporting methods is both time consuming and resource intensive as a typical OR&A system monitors approximately 2,500 items during the project.

Obviously, maintaining a spreadsheet of such magnitude would be onerous enough, but to record and catalogue comments and progress at this level would be prohibitive and definitely not cost effective. Similarly, producing detailed reports on a weekly and monthly basis would exacerbate the problems.

1.4.2 Inherent Assurance

The 'it will be fine' perception provided by inherent assurance is bolstered by the views of many 'old hands' in Oil & Gas projects who will tell everyone that they have done it all before, many times over and have survived to tell the tale.

Since the Piper Alpha disaster in July 1988 the authorities have been prompted to legislate to prevent the recurrence of major incidents, but this has simply served to perpetuate the false impression of perceived assurance. History however, tells a different story:

- Gannet Alpha platform, North Sea, UK, **GAS LEAK**, Feb 2012
- North Apoi platform, Nigeria, **FIRE**, January 2012
- Deepwater Horizon, Gulf of Mexico, **FIRE & OIL LEAK**, Apr 2010
- West Atlas Jack-Up, Montara Field, Australia, **FIRE**, August 2009
- BP Texas City Refinery, Texas, USA, **FIRE**, March 2005
- Petrobras P-36, Brazil, **EXPLOSION, SINKING**, March 2001

Unfortunately, many people lost their lives or were seriously injured in almost all of these incidents, yet they continue to happen with alarming frequency, on Oil & Gas projects.

1.4.3 Real Assurance

Providing 'Real' assurance that an asset has achieved a state of Readiness for Operations and that the organisation is capable of the safe, sustainable and environmentally responsible long term running of an asset is a complex issue.

This can only be achieved using a rigorous process to record the necessary parameters that must be documented, examined and evaluated such that the level of Readiness for Operations can be determined.

It is not sufficient to assume that, because Project Management records the progress of a project or the quality of certain items that the stakeholder can in any way be assured that the asset will be in a state of Readiness for Operations by the due date.

It is often the case that the issues which affect the operability, maintainability and availability of a 'completed' asset do not become apparent until the long term operations activities commence.

Most projects are both time and cost constrained. This means that any suggestion that an additional process is required during the Project Management Process is often unwelcome, to say the least.

1.4.4 Achieving Real Assurance

This book is intended to demonstrate that there is a cost-effective way to provide 'Real' assurance.

Those who have implemented an OR&A process in the way described in this book have found it adds value, often provides a net cost saving and frequently improves the performance of the asset in the long term.

However, some have begun the process by creating the framework of a system, subsequently failing to ensure the robustness of that system or allowing the system to 'deteriorate' due to a lack of basic maintenance.

It is essential to understand what is required from an OR&A system before building the basic structure.

For this, we use the generic term 'Operations Requirements' to describe the components needed to ensure that the operations team will be both capable of the safe and efficient operation of that asset, in a safe, sustainable and environmentally friendly manner and can also be assured of the Operability, Maintainability and Availability of the asset.

This is effectively the 'Target Value' for any OR&A system.

Once this 'Target Value' is decided, determining the status of a project against this target value requires a robust structure identifying the key values to be measured which must be populated from Corporate information (Directives, Regulations, Rules, Processes, Instructions, Codes of Practice etc.), Best Practices, Lessons Learned, normative and legislative requirements

This will be a significant undertaking and will need to be done by Senior, Qualified and Experienced OR&A practitioners, however, once completed, the process can be re-used over and over again on every project the company undertakes.

The system can then provide 'real-time' assurance of the Readiness for Operations at any point in the project. Moreover, once the system is built it can also be used on subsequent projects.

Note: Real time assurance requires the use of the OR&A desktop application from OR&A Ltd.

2 Mega-Projects or Mega-Failures

'The difference between failure and success is the difference between doing a thing nearly right and doing a thing exactly right'

Edward Simmons, 18th Century Engineer.

2.1 Identifying the Challenges

Over the last 20 years, a large number of high value Oil & Gas projects moved into the operational phase and as they did so, many were already exhibiting major problems (particularly operational issues).

This was recognised and highlighted in reports from the leading project performance monitoring companies, (including IPA Inc. and Boos, Allen Hamilton).

The issue is serious, not least because of the magnitude of finance involved, meaning that stakeholders is such projects cannot wait until the project is declared as complete, and handover to the operations team imminent, to discover that something has been omitted or overlooked which would subsequently prevent sustained operation of the asset to the design intent.

In his publications on the subject, Edward W. Merrow, founder of IPA Inc., asks four pertinent questions intended to identify failed megaprojects*,

- Will it be built and started up without causing harm?
- Will the total cost be within 25% of the budget estimate?
- Will it be completed on schedule (less than 25% late)?
- Will it deliver what was promised (the design intent)?

Data from over 300 projects indicated that **65%** of them **failed** on one or more of these indicators.

- A project is considered to be a megaproject if the total estimated project value exceeds US$ 100 million.

The issue dealt with by this book, particularly in respect of the operational issues and the use of the Operations Readiness & Assurance (OR&A) process, discusses much of what can be done to reduce, mitigate or avoid these problems.

2.2   Asset Operations Team

Involvement of representatives of the Asset Operations Team as early as possible in a project is the only way to ensure that critical decisions, especially those taken in the early stages of the project, can be validated by suitably qualified and experienced people.

By ensuring the involvement of senior, experienced and competent operations personnel from the earliest stages of the project, (applying the OR&A process), the asset owner can be confident that the key decisions which affect the integrity, operability, maintainability and sustainability of the asset are made with the end user in mind.

The OR&A Process encompasses the following key concepts:

- Front-End Loading (FEL) to ensure that key decisions with high cost implications are made before the cost of changes to those decisions becomes prohibitive.
- Operations Requirements specified from the outset, validated at each phase, including Operability, Maintainability and Availability of the asset.
- Incorporation of Lessons Learned and Best Practice examples from previous and similar projects to avoid many of the mistakes made in the pre-feed and early design phases, when practical operational experience is often overlooked.
- Flawless Project Delivery, where the concept of doing activities right first time, leading to a smooth start-up, steady ramp-up to the design parameters and sustained operation is applied.
- Life-Cycle Operating Costs (OPEX) and Total Cost of Ownership (TCO) to ensure the sustainability (& profitability) of the asset.
- Readiness of the Operations Organisation, including the Operations Team to operate and maintain the asset.

However, this can only be effectively applied using a structured process such as OR&A and a rigorous governance process to monitor compliance to the policies, standards, business processes and procedures required. Using the OR&A approach, the challenges identified can be significantly reduced, mitigated and often prevented from occurring at all.

2.3 Project Timeline

All standard project management techniques identify their timeline as a series of stages or phases. Typically, this timeline has a number of decision gates (usually reviews or audits) that control the progress of the project from one phase to the next.

In companies who manage this process in a 'Best Practice' fashion, the transition from phase to subsequent phase is controlled by a decision mechanism such as a gate review or audit.

Projects usually start with a designated Gate '0' or some form of formal project initiation document and end at a formal close-out in the final phase (usually early in the Operate phase) such as a Gate or Audit.

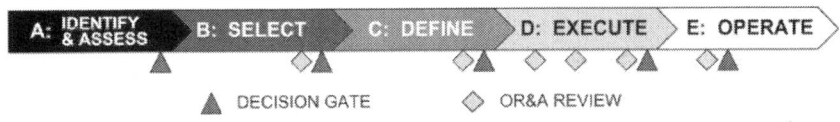

▲ DECISION GATE   ◆ OR&A REVIEW

Figure 3 -   Project Phases & Gates

The phases are often given (slightly) different names by different companies, but each has similar purposes:

- **IDENTIFY & ASSESS**, where the initial assessment of an opportunity is carried out and basic concepts for development are created;
- **SELECT**, where a preferred concept for further development is selected,
- **DEFINE**, where the details of the development are defined in support of the investment decision to design and construct an asset,
- **EXECUTE**, where a fully detailed design is performed and the asset is constructed, commissioned and started-up,
- **OPERATE**, where the asset is performance tested to the design intent and handed over to the owner/operator.

The OR&A process also uses this project timeline, synchronising the timing of activities needed to complete specific OR&A activities, Operations Requirements (and OR&A Reviews) in each phase, to the project timeline and the Integrated Project Plan (IPP).

2.4   Front End Loading (FEL)

The concept of Front End Loading (FEL) is very important in any project as it relates specifically to those decisions made in the early project phases.

During the early phases of a project, or the 'Front End', the extent of the project work is usually limited to the process of identifying appropriate concepts and selection of a preferred option for further development. The work is carried out by the project team (and specific contributors) and at this stage, is purely conceptual in nature.

What this means is that decisions may be refined, modified or changed relatively easily because the influence over the project is extensive and the cost of the changes is relatively inexpensive to implement.

In the later stages, when detail engineering design is involved, equipment is being ordered or purchased and site work is beginning, the ability to influence the design and make changes is reduced while the cost of making those changes is high (especially if construction is at an advanced stage).

It makes very good sense to ensure that the 'Front End' activities are as comprehensive (or 'Loaded') such that the key decisions are made, confirmed and validated in the early phases.

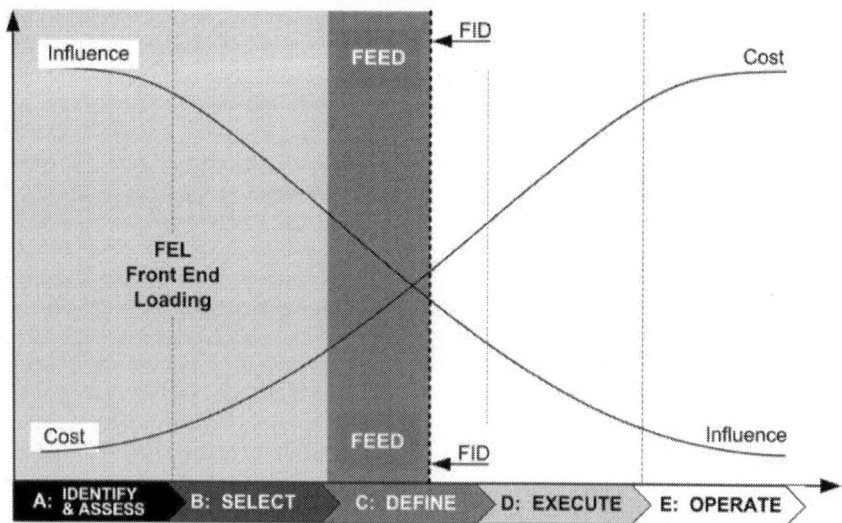

Figure 4 -   Influence vs. Cost in Project Phases

As illustrated by the graph, by the time the Front-End Engineering Design (FEED) is completed and the Financial Investment Decision (FID) is taken, any changes will be more difficult and more expensive to implement, making it imperative to get things right from the outset.

2.5 Business Case for Using OR&A

Invariably, due to the way the project process is implemented in most oil & gas exploration and production companies, there is a greater emphasis on the financial aspects of any project.

So, when an opportunity for a project is identified, the decision to move forward is based primarily on whether it will be a financial success.

Obviously, the parameters used to determine that success at this early point in a project are based on estimates of the cost to build the asset and the time taken to do so, compared to what the completed asset will deliver in terms of financial returns, how long it will take to repay the original financial investment and the estimates of productivity in the longer term.

In project terms, this leads to a Financial Investment Decision (FID) that determines whether the owner should proceed with the development.

The FID is based on the data prepared by the development framing team, as described in the previous paragraph, often referred to as the Design Intent or the Promises Made at FID.

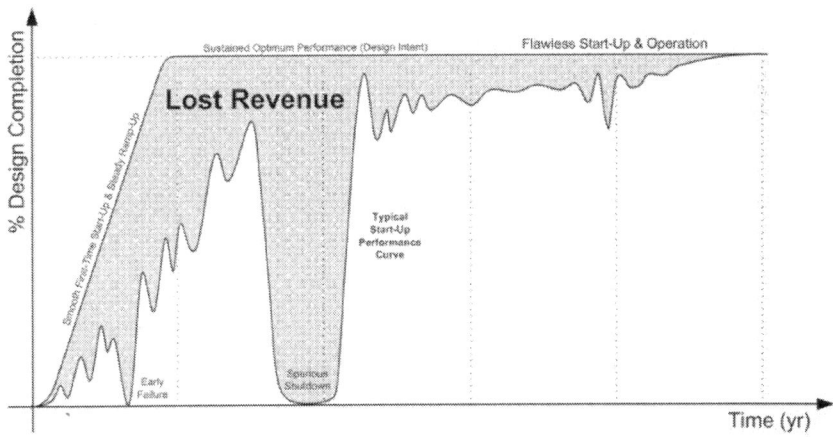

Figure 5 -   Design Intent vs. Typical Performance

However, as the graph illustrates, the financial return is based on sustained output over a period of time, not just at start-up, making operability, maintainability and availability key components in achieving that Design Intent (or fulfilling the Promises Made at FID).

According to recent project monitoring statistics, in a significantly high percentage of projects, the concept of the Operator 'being ready to operate' the asset is not high on the priority list of most Project Managers and their respective teams.

Most projects are Cost and Schedule driven, measured by physical (mechanical) completion over a period of time, usually represented in terms of the Value of Work Done (VoWD) and Earned Value (EV) as measured against the project schedule.

Typically, (but more especially on failed projects), this leads to a point where the schedule has ended, the budget is spent and yet the work is not fully completed. Consequently, the Project Managers involved would regard it as normal to handover an asset that is 'mostly' complete, but with a very long 'punch-list' of incomplete items.

Although in the majority of cases hydrocarbons are safely introduced and the asset start-up completed without incident, financial pressures (cost overruns) on the Project often mean that there is little or no funding to complete the 'punch-list' items.

The additional work, further delays and extra cost for the asset operator before the asset can be considered to be fully completed if a significant drain on the OPEX in the early OPERATE phase of an asset and affects the projected financial viability of the asset.

The aim of the OR&A process is to ensure that the Operations Organisation is 'Ready to Operate' the asset and achieve a flawless start-up and a smooth ramp-up to sustained steady state operation of the asset at the design parameters (fulfilling the Promises Made at FID).

Typically, when the priority in these projects is not given to the sustainable operation of the asset, it is often the case that although the start-up was effected on or before the schedule date, operability issues mean that sustained, steady operation is difficult to achieve and any aspiration of Operations Excellence is remote (to say the least).

2.6 CAPEX vs. OPEX

The economics of the longer term operation of the asset are obviously less important to the designer and constructor of an asset and are often disregarded by the design & construction teams as being largely irrelevant.

This may seem to be acceptable in the context of the limited project timescale, but to the Asset Owner/Operator failure to consider such issues can make the difference between a profitable asset and a loss making venture.

During the design phase, a preliminary estimate of the running costs of an asset, also known as Operating Expenditure (OPEX), is estimated at around 10% of the total value of the project, or Capital Expenditure (CAPEX).

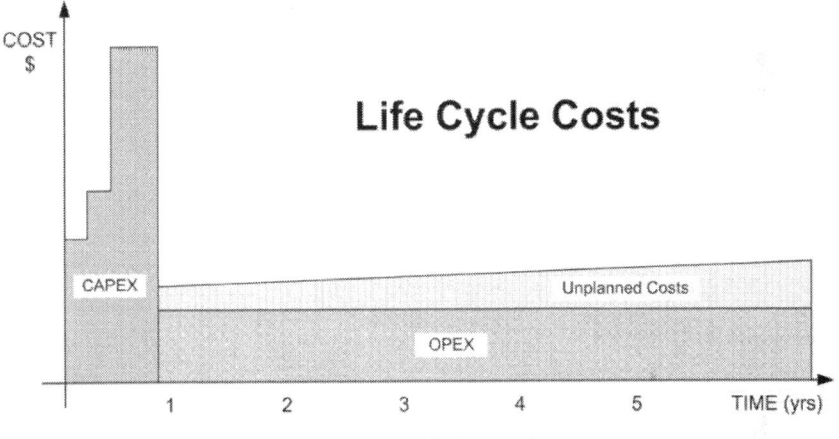

Figure 6 -   Life Cycle Costs

However, over a typical design life of an asset of 25-30 years, that 10% per annum OPEX figure can easily exceed 300% of that original CAPEX value over the life of the asset, exacerbated by any inflationary effects.

When other unplanned costs such as breakdowns, abnormal process conditions of the non-availability of spares are added in to these equation, the value of making the right decisions and including the Operations Requirements, (especially during Front End Loading activities), becomes immediately apparent.

2.7 Introducing OR&A

Operations Readiness & Assurance (or OR&A as it is commonly referred to) is a process used mainly in Oil & Gas projects to ensure the adequate preparation of the asset organisation for the eventual operation of an asset.

It is employed throughout a project, from the earliest possible moment, to ensure that the Asset Operations Team is involved in making and validating key decisions that affect the operability and maintainability (and therefore sustainability and profitability) of the asset.

However, delays in implementing OR&A are often experienced when the Project Owner does not fully understand the purpose and benefits of OR&A. To delay implementing OR&A until the Final Investment Decision is taken (as many Companies presently do) makes the benefits of using OR&A more difficult and often more expensive to achieve (see Front End Loading (FEL)).

The OR&A process is intended to ensure the appropriate preparations for the operations phase are implemented in a timely manner.

The OR&A methodology also includes provision of on-going assurance that those preparations will be completed at the point of handover of the completed asset from the Project Team to the Asset Operations Team.

Additional Work Due To OR&A Activities

If the Project Team performs all of its work on schedule and to the correct standards of quality and completeness, the OR&A Process is effectively invisible.

In this instance, OR&A Process is only required to monitor the project work specifically related to the Operability and Maintainability of the asset under development, to ensure the preparedness of that asset and the supporting organisation at the point of handover.

Put simply, if the Project Team does its job properly, there is no extra work involved other than the work carried out by the OR&A Team to implement the OR&A Process.

2.8 The New Car Analogy

2.8.1 Similarities in the Processes

Sometimes, when introducing new concepts, it is useful to do so using a set of circumstances more familiar to the user by way of an illustrative analogy.

The analogy we have found to illicit the best response from operations and project personnel with little or no experience of the concept of OR&A is that of choosing, purchasing and taking delivery of a new car.

Describing the selection and decision processes that they would use in the car purchase, then identifying the similarities between that process and implementing OR&A on a project, this allows us to provide a logical introduction of the terminology and thought processes that are used in the OR&A process, for example:

2.8.2 Operations Requirements

The first step in buying a car is deciding what type of car you would like and how much you are prepared to pay for it:

- What type of car do you need?
    - Sports Car,
    - Saloon Car (Sedan),
    - Estate Car (Wagon),
    - Van,
    - Pick-up Truck.
- How much will it cost?
    - List Price, Taxes & Delivery Costs (CAPEX),
    - Registration & Insurance Costs (OPEX),
    - Running Costs (OPEX).

This could be compared to the Opportunity Framing process where a Project Owner decides how to develop an opportunity, identifies and compares a number of options and prepares a basic cost estimate.

IDENTIFY & ASSESS

At this stage in the project, in the IDENTIFY & ASSESS phase, the decisions made provide the outline of the eventual asset (the type of car) and an initial estimate of CAPEX costs (list price) and OPEX (running costs) the asset (car).

Even at this level, the car analogy demonstrates striking similarities to an Oil & Gas project, even to the point of the typical CAPEX vs. OPEX considerations.

SELECT

By the time the car analogy moves to the next stage (the SELECT phase), a number of high level options are being considered, once we have a body style in mind, we select the preferred option.

If the reason for needing a car is to move a lot of luggage around with us, we have identified a basic Operations Requirement, so the choice is likely to be from an Estate Car (Wagon) or a Van or a Pick-Up Truck.

Refining that choice might include considering the size and weight of the items of luggage and how much our selected vehicle will need to carry.

By the end of the SELECT phase, we will have a preferred option for further development, with a better idea of what it will cost (refined CAPEX and OPEX estimates).

DEFINE

When we reach the DEFINE phase, we have chosen a type of vehicle and a basic specification, but we now need to go into more detail with our specification to ensure our car has all of the necessary attributes to satisfy our needs:

- Body Style,
    - Interior Trim (fabric/leather seats, aircon, ICE etc.)
    - Colour (plain or metallic paint, stripes, body kit etc.)
    - Exterior Trim (wheels, tinted glass etc.)
- Physical Size,
    - Passenger Capacity (no. of seats) & Load Capacity,
    - Operating Range (Fuel Tank capacity vs. estimated fuel consumption),
- Engine Type/Capacity,
- Fuel Type (emissions?),
- Maintenance Requirements,
- Availability,
- Other Running Costs (insurance, road tax, consumables).

At this stage we move through Front End Engineering Design (FEED), making more detailed decisions about our Operations Requirements, refining the Basis for Design (BfD) such that it reflects our specifications for the end product (asset/car).

Armed with our BfD and a more refined cost estimate, we can look for a supplier, we can visit the car dealerships to match our requirements to what they have on offer (or we can send the Invitation to Tender (ITT) to a number of prospective EPC contractors).

EXECUTE

All projects reach a point where a decision has to be made to spend money (to invest the finance to achieve the project objectives). In any case, whether we are talking about an oil & gas project or just buying a car, this will involve partners, stakeholders and possibly financial investment from other sources.

In oil & gas terms, this is called the Financial Investment Decision (FID) and could be considered as very similar to obtaining the wife's permission to purchase the car!

At this stage of the project, the real volume of work begins in establishing the fine detail of what is required.

- Checking, verifying and confirming that the design choices will be fulfilled,
- Researching the detail from the vendors and supplier, checking performance details and guarantees, (dealer literature, motoring press, road test reports etc.).
- Seeking assurance that the right choices have been made is all part of the Due Diligence process.

This is normal before committing to a large financial outlay and should be applied equally to both buying the family car and a $1,000 million mega-project.

There will not many families who delegate this work to an agent of some kind, but large Exploration & Production (E&P) companies rarely have the manpower in-house to do this work and therefore engage the services of an Engineer, Procure & Construct (EPC) Contractor Company to do it on their behalf.

Although the E&P Company awards an EPC contract to a contractor company, they will still want to oversee the work to confirm that the contractor is building what they want, to the necessary standard and quality and to ensure it will be delivered on time.

Once the order is placed for the car, the typical family has no involvement in the design; construction and testing of a vehicle, but will still monitor progress against the promised delivery date.

They will also be involved if the manufacturer needs to change the specifications of what they have ordered, to ensure that subsequent changes and the costs involved, do not cause detrimental changes to the expected outcome and the car will still do what it was intended to do.

Where the EXECUTE phase activities are also similar is in preparing the Operations Organisation (the family) to ensure they are ready to operate the asset (car) at the point of handover (the OPERATE phase).

There are several aspects to consider in preparing for the OPERATE phase:

- Do all users have the appropriate driving permit
  (is competency established)?
- Does anyone need training?
  (Driving Lessons, Test, Refresher Training etc.)
- Does the family have the correct insurance?
- Is the road tax paid (or does it need to be)?
- Is the site ready for operations?
  (where will it be parked, does it fit in the garage)
- Does the family have the necessary tools and equipment for first line maintenance?
  (Who will wash the car, check the oil, kick the tyres etc.)
- How will the car be maintained?
  (if it will be dealer maintained, where is the dealership)
- What about breakdowns?
  (Do we have auto-club membership? what warranty do we have? what about spare parts?)

In a typical mega-project, OR&A will be involved in every phase to ensure all these questions are answered to provide assurance to the Project Owner.

Key members of the Operations Team will usually be dispatched to the various vendor supplier premises to participate in Factory Acceptance Testing (FAT), Site Acceptance Testing (SAT) and embedded in the Commissioning and Start-Up Team to give them the maximum hands-on experience with the asset equipment prior to the OPERATE phase.

OPERATE

After start-up, the Operations Team will be involved in the ramp-up of production to the design intent parameters and participate in Performance Testing activities to confirm the asset achieves the sustained performance promised at FID.

As with Oil & Gas assets, after delivery of the vehicle (Handover & Acceptance), the owner is responsible for the safe operation and maintenance of the vehicle both during and beyond the Warranty Period.

Other aspects considered by OR&A on behalf of the asset owner should be:

- Succession Planning, to address workforce changes and to maintain the level of competence of the workforce.

- Decommissioning, it is rapidly becoming the norm for projects to address decommissioning of an asset during the design process, (to ensure that an asset can be safely dismantled and removed and that the site can be returned to its original condition relatively easily).

SUMMARY

From this analogy it can be plainly seen that Operations Requirements must be considered from the earliest stage of any project to ensure the outcome, (whether it be a new car or an oil & gas asset), meets the initial requirements.

OR&A will ensure that any changes made to those requirements along the way does not significantly change the initial design premise or detrimentally affect the longer term performance (in terms of throughput, availability, maintainability, operability or cost).

Finally, frequently checks during the process will ensure that when the asset (new car) is delivered, it can be driven it in the way initially intended. In other words, the Operations Organisation is fully equipped to operate the asset to the original design parameters in a safe, sustainable and environmentally friendly manner for the intended design life of the asset.

This is the main purpose of the Assurance component of OR&A throughout the process. It is also to identify and monitor the Operations Requirements and anything else that may impact the asset after handover and bring this to the attention of the Project Owner (or his nominated deputy) in a timely manner such that effective remedial action can be taken before the cost of the remedy threatens the outcome of the project (or the long term operation or viability of the asset).

3 Introduction to OR&A

3.1 Definition of OR&A

Operations Readiness describes the process of preparing the custodians (owner/operator) of an asset under construction (and the supporting organisation) such that, at the point of delivery/handover, that organisation is *Ready to Operate*, i.e. fully prepared to assume ownership of the asset, accepts responsibility for, (and is capable of), performing the safe and efficient operation of that asset, in a safe, sustainable and environmentally friendly manner.

By default, this includes participation in the Project activities to ensure the needs of the Asset Operations Team (to enable them to do this) are recognised and addressed during the development of the asset.

Assurance, in this same context, refers to the act of providing assurance to the various stakeholders in a project or venture that both their asset and supporting organisation is in a state of *Readiness to Operate*, (or that it will be, by the time acceptance/handover occurs).

Assurance plays a key role in the risk management process and decisions often made without considering all the factors are made because of a lack of information or appear too complex to understand.

The OR&A process provides an analysis of the risk to Operations Readiness to allow more informed decisions to be made.

OR&A is a continuous process that is implemented from the earliest point in a project, ending after a successful handover to the asset custodian is achieved.

3.2 OR&A Pre-Requisites

3.2.1 Management Commitment

The key to successful implementation of the OR&A process is the commitment of management to the principles of OR&A, backed up by the allocation of adequate funds and resources.

If this Corporate/Management buy-in is not demonstrated from the outset and maintained throughout the implementation of the process to its conclusion, the benefits of the OR&A process will never be fully realised.

3.2.2 Management Structure

For any project to be managed a Control Framework (Management Structure) is required, this usually takes the form of Directives, Regulations, Rules, Policies and Procedures.

This Control Framework is normally expressed as a defined Corporate Management System with a formal hierarchy, supported by a clearly defined Designated Authority Matrix.

If such a robust structure exists, implementing OR&A is much easier to achieve.

3.2.3 OR&A Team Competencies/Experience

To implement the OR&A Process on any project requires senior, experienced and qualified practitioners to ensure that implementation to the appropriate quality and standard.

The ability (and authority) to apply the necessary rigour to the project requires that the OR&A Manager/Lead to be of an appropriate level of seniority in the Owner/Operator Organisation.

	Projects Experience	Operations Experience	OR&A Experience	Seniority
Senior Advisor/Expert (Technical Authority)	10 yrs	10 yrs	10 yrs	Advisor
Lead OR&A Practitioner (Project OR&A Lead)	5 yrs	5 yrs	5yrs	Expert
Senior OR&A Practitioner (Discipline Lead)	5 yrs	5 yrs	2 yrs	Highly Experienced
OR&A Practitioner (Team Member)	2 yrs	2 yrs	0 yrs	Experienced

Figure 7 - Typical OR&A Competency Matrix

All OR&A personnel must be embedded in the project organisation and will report directly to the Project Owner or his nominated representative in the asset organisation (not the Project Manager).

The OR&A Manager/Lead will also have a functional reporting link to the corporate organisation in respect of the Technical Authority for the OR&A Process.

4 Developing an OR&A Capability

4.1 Management of the OR&A Process

An OR&A system is a robust process to implement and control the principles and practices of OR&A in any project. As with any true management system, the system applies the principles of ISO 9000 to control the process.

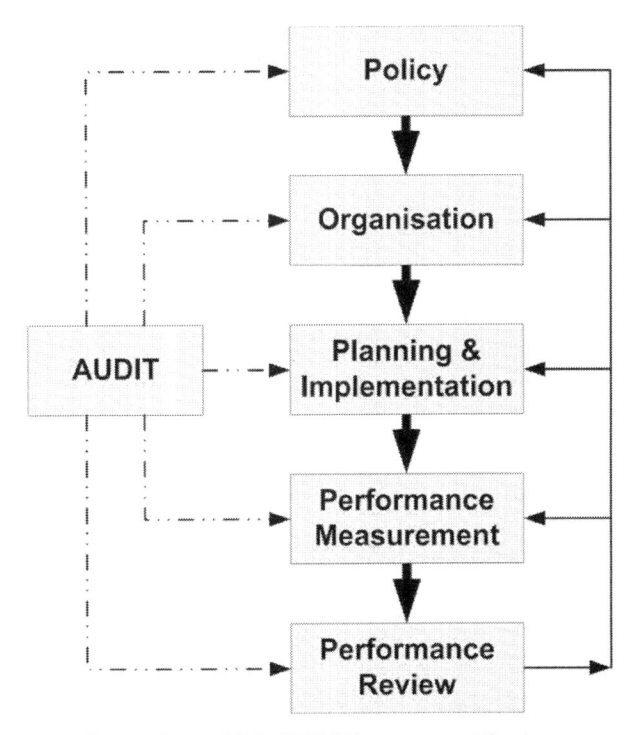

Figure 8 - ISO 9000 Management System

4.1.1 Policy

The Corporate Management System adopted by most client systems is modelled on the ISO 9000 process. At this level, provided that this is the case, the policies, plans, and business processes of the client provide the structure both reflected and employed in the OR&A System.

4.1.2 Organisation

The organisational structure used by client organisations is also reflected in the organisational structure of the OR&A Process to enhance the synergies between the two systems.

4.1.3 Planning & Implementation

The OR&A Database contains key content from the OR&A Information System, aligned with the project using the project phases.

The database facilitates the generation of various output used to implement OR&A on a project, including an OR&A Plan and an OR&A Schedule.

The OR&A schedule becomes part of the Integrated Project Plan (IPP), aligning the OR&A activities with the project timeline and milestones.

4.1.4 Performance Measurement

The OR&A Database is used to record progress of the relevant activities and tasks on a day-to-day basis, together with other comments and data to support that information.

4.1.5 Performance Review

The OR&A Database is also used to generate the various reports, such as the performance dashboard which provide information in a pre-determined format by which the implementation of the OR&A Process and the performance of OR&A Team can be easily and accurately evaluated.

4.1.6 Audit

Auditing the implementation of the OR&A Process, utilising the same tool that is used by OR&A Engineers to monitor and record progress and to generate reports, is very easy to do in a structured manner.

The dashboard output serves as a very effective auditing tool which both speeds up the audit process and increases the accuracy of the audit process.

4.2 The OR&A Information Structure

The structure of the OR&A Structure described in this book should appear familiar when compared to the generic Corporate Information Structure adopted by most corporate entities.

4.2.1 Corporate Information Structure

The typical Corporate Information Structure can usually be visualised as a pyramid with a discrete number of levels where the focus on any specific component part of the system is increased as one moves down the pyramid through each successive lower level.

CEO
Board of Directors

GOVERNANCE
Normative Documents,
Rules & Regulations

BUSINESS PROCESSES
Processes, Systems, Instructions, Procedures, Guidelines

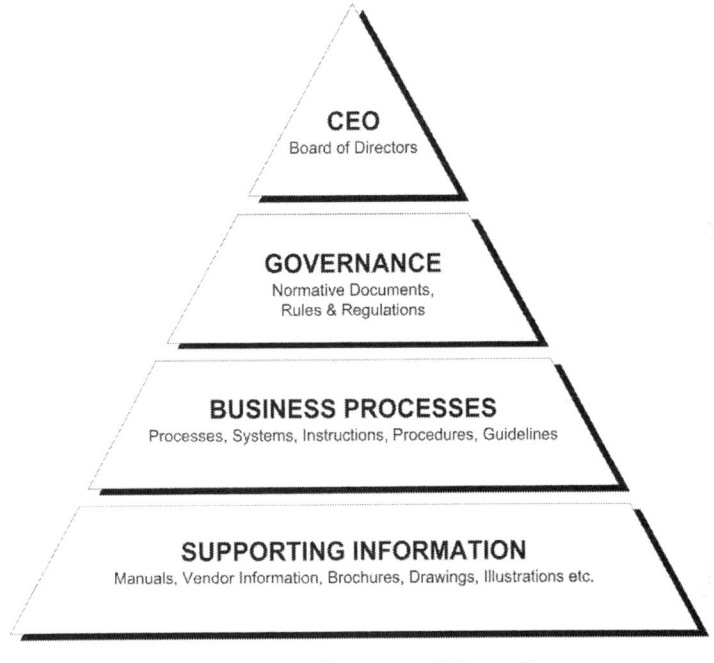

SUPPORTING INFORMATION
Manuals, Vendor Information, Brochures, Drawings, Illustrations etc.

Figure 9 -   Corporate Hierarchy

At the top of the pyramid, the policy, strategy, mission and objectives of the organisation is set and implemented by the corporate management team, (i.e. the CEO).

At the next level down, legislative, normative, regulatory documents and rules specify the permitted parameters within which the business activities of the organisation may be carried out.

This is the origin of the governance processes to ensure the organisations activities comply with the relevant laws.

At the next level down, the company business processes, guidelines, procedures and instructions specify the activities to be carried out to perform the work of the company. These are usually written and comply with the relevance governance requirements.

At the lowest level of the pyramid, there is usually a library of some sort which contains the relevant records, data and reference information required by the workforce in completing the work to the required quality, standard and completeness.

The hierarchy of the information, particularly in respect of governance and the business processes represents a structured Control Framework for the activities of the organisation, supported by the Knowledge Base held in the library. The Corporate Control Framework of the client company or asset operator will therefore usually include, though should not limited to, the following:

- Corporate Management System
- International, National and Local Legislation
- Corporate Directives, Rules & Regulations
- HSE Management System
- Capital Project Management System
- Operations Management System
- Business Processes, Procedures & Instructions
- Knowledge Base, Records, Data and References

4.2.2 OR&A Information Structure

The information structure of the OR&A Process is similar to a typical Corporate Information Structure, as illustrated below:

OR&A SYSTEM

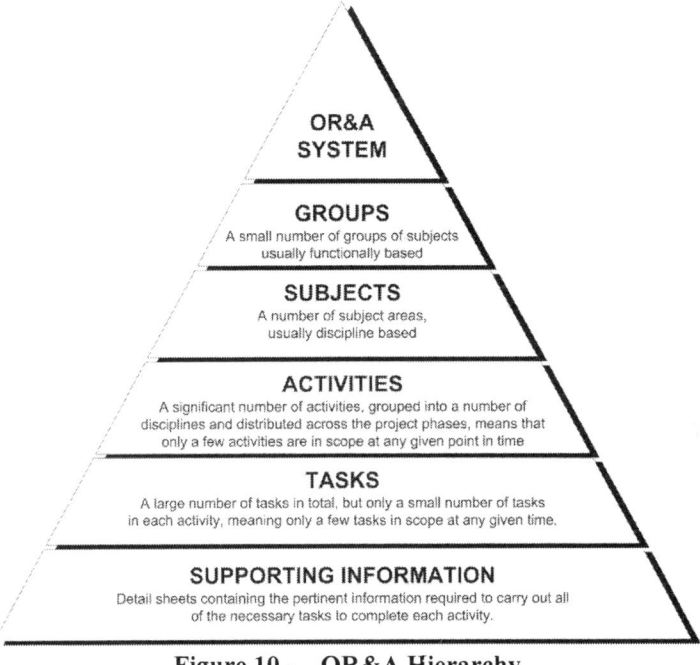

GROUPS
A small number of groups of subjects usually functionally based

SUBJECTS
A number of subject areas, usually discipline based

ACTIVITIES
A significant number of activities, grouped into a number of disciplines and distributed across the project phases, means that only a few activities are in scope at any given point in time

TASKS
A large number of tasks in total, but only a small number of tasks in each activity, meaning only a few tasks in scope at any given time.

SUPPORTING INFORMATION
Detail sheets containing the pertinent information required to carry out all of the necessary tasks to complete each activity.

Figure 10 -   OR&A Hierarchy

The OR&A Information Structure can be similarly visualised as a pyramid with a discrete number of levels where the focus on any specific component part of the system is increased as one moves down the pyramid through each successive lower level:

- Level 0 - OR&A System
- Level 1 - OR&A Groups
- Level 2 - OR&A Subjects
- Level 3 - OR&A Activities
- Level 4 - OR&A Tasks
- Level 5 - Reference Library

5 OR&A System Structure

5.1 The Work Breakdown Structure (WBS)

A Work Breakdown Structure (WBS) is a complete listing of every activity, task or deliverable required to complete the work within the planned timescale of the project.

A WBS is a logical decomposition of the total work required, grouping the whole into a number of subjects, each of which comprises a number of packages (Activities) and smaller parts of each package (Tasks). This is done by following a few 'Rules of Thumb' to ensure consistency:

- No task longer than a man/week (the first line reporting cycle);
- No Activity with a duration longer than a month
  (or the basic project reporting cycle);
- Common Sense moderation applied to the above rules.

The Work Breakdown Structure does not specify how or when the work should be carried out (or in what order it should be done). It simply lists the entire work required, to complete the OR&A activities in a project, in a logical manner (or structure).

In terms of creating an OR&A System, this is the first step of the approach used to construct the OR&A Information Structure.

5.1.1 Generic OR&A Groups

The highest level of the WBS decomposes the work required to complete the OR&A activities into a number of functional groups or categories and typically, these groups will have similar titles to those listed below:

- OR&A Management
- Operations Management
- Health, Safety and Environmental Management
- Operations Organisation & Competence
- Commissioning, Start-Up

These groups are directly related to the involvement of the Asset Operations Team in key components of the project, which have an effect upon the Operation of the completed asset.

5.1.2 OR&A Subjects

Each OR&A Group can be further divided into a number of discrete elements identified in this book as 'Subjects', (but could also be described as generic disciplines, topics or areas of influence within the project scope), providing a further level of granularity to the developing OR&A structure.

Subjects within the remit of OR&A will typically include the following:

1.0 OR&A Management
  1.01 OR&A Management
  1.02 Cost Estimates (OPEX) & Management
  1.03 Quality Management
  1.04 Licences, Permits & Permissions Management
  1.05 Flawless Project Delivery

2.0 Operations Management
  2.01 Operations Management & Integrated Planning
  2.02 Production Operations
  2.03 Production Chemistry (& Flow Assurance)
  2.04 Wells (Design, Operation & Management)
  2.05 Infield & Export Pipelines & Manifolds
  2.06 Sub-Sea Systems (Design, Ops & Management)
  2.07 Maintenance & Integrity Management
  2.08 Process Automation & Control Systems
  2.09 Geomatics, Logistics & Infrastructure
  2.10 Materials Procurement & Materials Management
  2.11 IM, IT & Communications Systems
  2.12 Contracted Services (Strategy & Management)

3.0 Operations Organisation & Competency
  3.01 Operations Organisation
  3.02 Operations Training & Competency

4.0 Health, Safety, Security & Environmental Management
  4.01 Health, Safety & Environmental (HSE) Management
  4.02 Security Management
  4.03 Community Relations & Sustainability
  4.04 Risk Management

5.0 Commissioning & Start-Up (CSU) Management
  5.01 Commissioning & Start-Up (CSU)
  5.02 Asset Handover & Acceptance

5.1.3 Activities

The project OR&A Team must methodically address each Subject in a structured manner, carrying out a number of activities in each project phase.

Some of these activities will need to be carried out in sequence and some will need to be cross referenced to other activities (perhaps in different subject areas).

However, all of these activities must be completed in each project phase in order to complete the work required to ensure that the Asset will be 'Ready to Operate' at the point of handover.

This means that the OR&A Team must be made aware of a number of specific activities which must be performed and specific deliverables that must be produced at appropriate phases, stages or key points in the project timeline.

As previously mentioned, the OR&A process described in this book uses the typical project timeline, synchronising the timing of activities needed to complete the OR&A activities, addressing the Operations Requirements and OR&A Reviews in each phase, in relation to the project timeline and the Integrated Project Plan (IPP).

The OR&A process also uses this project timeline, aligning the activities needed to complete specific activities, operations requirements (and OR&A Reviews) in each phase with the project milestones (illustrated below).

A: IDENTIFY & ASSESS B: SELECT C: DEFINE D: EXECUTE E: OPERATE

▲ DECISION GATE   ◇ OR&A REVIEW

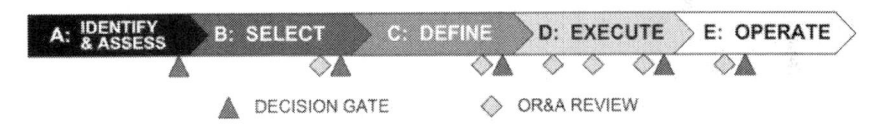

Figure 11 - Decision Gates & Reviews

Each phase is usually identified both by name, as in this book and each has been given a sequence letter (for reasons which will be explained in more detail later in this book), also in the OR&A system, each phase is also colour coded (using colours of the spectrum, in sequence).

5.1.4

5.1.5 Tasks

Simply identifying the tasks in not sufficient to ensure the comprehensive implementation of OR&A in a project. Because the definition of what OR&A comprises is, to date, unclear to many OR&A practitioners, it is necessary for any comprehensive system to delve deeper into what is required.

The OR&A system described in this book advocates that each activity should be further decomposed to produce a list of tasks and to provide guidance and supporting information to enable those tasks to be completed to the necessary standard, quality and completeness required, by the most efficient method.

The OR&A System content, including the **Operations Requirements**, is extracted from the core business processes of the client company (asset owner/operator) and documented in their various rules, guidelines, instructions and procedures.

This should be complemented by further information sourced from international, national or regional regulations, rules and standards, from Lessons Learned and industry Best Practices or as is usually the case, a combination of all of these.

Each activity therefore comprises a number of tasks which, when each is completed will ensure that every aspect of the activity in completed.

5.1.6 Deliverables

Deliverable is the term used to describe the product, result, output or outcome of any discrete task, activity or series of activities. It is usually a tangible item such as a document or certificate, but could also be the achievement of a specific OR&A or Project milestone.

Deliverables from the OR&A process include, but are not limited to, the following:

- OR&A Deliverables
    - Management & Implementation of OR&A,
    - Provision of on-going assurance,
    - Operations and OR&A Documents,
    - Activities led by OR&A
    - Activities led by other disciplines

- Project Deliverables
    - OR&A Contribution
    - OR&A Participation (monitor, comment, review etc.)
    - OR&A Validation (checking, witnessing & approvals)

The key to ensuring these are applied consistently is to use a structured system which provides the OR&A engineer with the right information at the right level of detail and at the right time, avoiding information overload.

5.2 Populating the OR&A Structure

5.2.1 Delivery Maps

Within the system developed by OR&A Ltd., a delivery map is prepared for each Subject. It is a graphic representation of the activities which need to be carried to implement OR&A in each phase of a project.

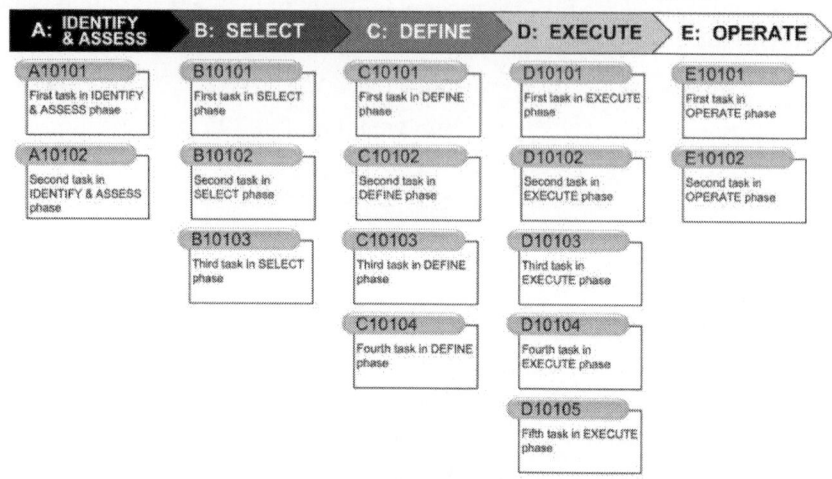

Figure 12 - Generic OR&A Delivery Map

5.2.2 Detail Sheets

To fully explain what is required to complete each activity, a Detail Sheet is prepared for every activity. It contains all of the necessary pertinent information to enable the OR&A practitioner to perform his duties to the required quality, standard and completeness.

These Detail Sheets are uniquely numbered with the same number assigned to the relevant Activity on the Delivery Map and as such can be 'placed' within the system at a glance.

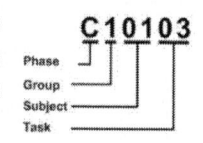

C 1 0 1 0 3
Phase
Group
Subject
Task

To further support the completion of the activity and associated tasks, references to targeted context specific information, including example and skeleton documents are included in the Detail Sheet content. This information is also included in the OR&A Library.

The most important job when creating an OR&A System is the population of the Detail Sheets. These are the source of the 'core' information used to implement OR&A, so it is vital that these are prepared by senior, experienced and competent OR&A Practitioners.

Fact, Not Fiction
The content of the Detail Sheets must be based on fact, whether that fact is sourced from the client company, international or national legislation, industry best practice or lessons learned. The Detail Sheet must identify the source of this information and, where possible, provide the appropriate references.

Clarity
The information on every Detail Sheet must always be clear, concise and unambiguous. There is no place for conjecture or opinion. Each task must specify only one purpose, because this is important to the monitoring activities inherent in the OR&A process/system.

Authority
During the implementation stage of OR&A, it is not unusual for the activities and tasks required to complete the implementation of OR&A to be challenged by the project team. This can only be defended by having the necessary references immediately to hand.

Ideally, as happens in many forward thinking Companies, a Corporate Directive of some sort will exist which demonstrates the management commitment to OR&A. It should specify that OR&A is a mandatory process for the project, and demonstrate this support by providing the necessary budget and resources.

Without this demonstration of that commitment, implementing OR&A becomes challenging and more costly as every challenge will require substantial effort to overcome.

5.2.3 Supporting Information (Library)

The Library pages on the system developed by OR&A Ltd. also contain a comprehensive listing of the reference information, example and skeleton documents to allow 'browsing' and informal search and retrieval of supporting documents. The library content is sorted into key types and the related OR&A Subjects to assist browsers to locate the exact documentation:

References / Miscellaneous Information
This refers to additional information such as International and Industry, Corporate Guidance, regulations, guidelines, policies, standards and other pertinent information.

Best Practice Examples
These are examples of documents taken from previous projects which are considered to be 'best in class' and suitable to be emulated in style, content and quality/standard.

Skeleton Documents
Documents which already contain the key content required but need to be 'tailored' to suit the current project.

These are more than just templates since they retain the generic document structure and headings and also contain guidance information intended to assist users in preparing documents of the necessary standard and quality to support the implementation of OR&A on a project.

6 OR&A System Content

6.1 OR&A Management (1.0)

6.1.1 OR&A Management (1.01)

As with any robust system, OR&A has a part of the system by which it is administrated. Subject 1.01 OR&A Management is the overarching subject which administrates and drives the implementation of OR&A on a project.

All systems start somewhere, OR&A begins at the earliest point in the project when the first OR&A resource is assigned a position in the project team and this should occur as soon as the Project Manager is given the green light to start the project, beginning with the creation of the Project Management Team.

This is the point at where the demonstration of management commitment to implementing OR&A on a project is most obvious. The speed and enthusiasm of the Project Owner and the Project Manager to identify, recruit and mobilise an OR&A Advisor/Manager to the Project Management Team will provide that demonstration.

This demonstrated commitment should be reinforced by the provision of the funds and resources to fully implement OR&A from the outset and if provided will maximise the benefit to the project.

This subject (OR&A Management) defines the scope of OR&A on the project and promotes the inclusion of **Operations Requirements** from the outset.

If the opportunity to benefit from the contribution of a senior, experienced and competent operations professional (the OR&A Manager/Advisor) to key project decisions is not taken at this stage, (most of which are high value decisions carrying a high cost of change should this be required in later phases), the project team could jeopardise the long term success and viability of the project.

Whilst this may not be immediately apparent, and possibly of little concern to the project community, (whose primary concern is to build and deliver the asset), it is of the utmost importance to the Project Owner and Asset Operator due to the effect it has on the Total Cost of Ownership (TCO) over the lifetime of the asset.

OR&A Management is primarily concerned with ensuring the implementation of the OR&A process, to making the necessary resources available to the project team and to ensuring they are both visible to and active within, that project team at all times.

OR&A Management as a subject promotes the inclusion of the OR&A Team in the following key activities and the provision of adequate resources to interact with the project team and to perform the necessary OR&A Activities to the necessary standard, quality and completeness.

These key activities include (but are not limited to) the following:

- Ensuring the Project team is aware of the OR&A process,
- Identification, recruitment and mobilisation of sufficient senior, qualified and competent OR&A personnel,
- Contribution of Operations expertise to the Project definition, scope and magnitude,
- Inclusion of OR&A in the Project Governance, Assurance Plans and Governance Activities,
- Participation of OR&A in Project Risk Management Activities,
- Inclusion of OR&A activities in the Integrated Project Plan,
- Inclusion of OR&A activities in project reviews, health checks audits etc.,
- Influence in the Opportunity Framing, Concept Selection and other decision processes during the project,
- Inclusion of appropriate Lessons Learned and Best Practices in Project decisions,
- Influence over the Basis for Design (BfD) and Invitation to Tender (ITT) documents to include **Operations Requirements** and address Total Cost of Ownership (TCO) issues.
- Control/timing/execution of Operations Readiness Reviews and Lessons Learned workshops (core components of OR&A).

As previously mentioned, in this phase, key decisions are made which can seriously affect the Operability and Maintainability of the asset.

Failure to address OR&A activities in a timely manner will risk missing opportunities to include Operations expertise at each stage of the project and overlook the contribution of the OR&A team to key decisions.

Critical Factors in the early inclusion of OR&A in a project may include the awareness of the Project Owner and/or the Project Manager of the OR&A Process resulting in delayed inclusion of the OR&A process.

Another more common failure is to assign OR&A resources without the necessary seniority, experience and competence, resulting in a lack of influence and reduced effectiveness of the OR&A process.

Typical Deliverables/Outputs from this Subject include:

- OR&A Plan - aligned with Integrated Project Plan
- Detailed OR&A Resources Plan
- Contribution to Project Assurance Plan
- Contribution to Project Assurance Plan
- Contribution to Opportunity Framing Process
- Contribution to Concept Selection Process
- Participation in Project Governance Activities
- Participation in Project Assurance Activities
- Participation in Project Risk Management Activities
- Contribution to Basis for Design
- Contribution to Invitation to Tender
- Inclusion of Lessons Learned & Best Practices
- Collection of new potential Lessons Learned

Basis for Design (BfD)
With all projects, the Facilities engineers and the project team will develop a Basis for Design document.

The Basis of Design is a functional performance specification for the plant and is used to provide key information to any prospective bidder for the construction of the asset.

The Basis of Design should specify, as a minimum:

- Well-stream composition (including produced water),
- Production profiles (from simulations),
- Production capacity (oil, gas and water),
- Product quality specifications,
- Environmental targets (e.g. Emissions),
- Flare capacity,
- Re-injection capacity (for gas and/or water),
- Product details,
- Facility life,
- Safety targets,
- Plant reliability/availability targets,
- Target operating costs,
- Approximate manning levels,
- Critical operability and maintainability factors,
- Block Flow Diagrams (BFD) for the process,
- Heat and material balance sheets for the process,
- Process data sheets for all major equipment.

6.1.2 Cost Estimates (OPEX) & Management (1.02)

The OR&A process only concerns itself with to separate cost related components (Cost Estimates) in respect of the project, these are:

- Preparation and refinement of cost estimates for the OPERATE phase activities of the Asset Team at each stage of the project, also known as OPEX,
- Preparation of cost estimates for the OR&A activities in each phase of the project (for OR&A budget/planning purposes).

OPEX

The primary purpose this subject is to develop a robust Life-Cycle Operating Expenditure (OPEX) cost estimate for the proposed asset in time for the Financial Investment Decision (FID) to be made at the end of the DEFINE phase and to allow detailed budgets to be prepared for the OPERATE phase.

Operating Cost (OPEX) Estimates usually begin, in the early project phases, with a high level estimate (for comparison) usually approximated somewhere between 2% - 10% of the Estimated Capital Expenditure (CAPEX) of the Project.

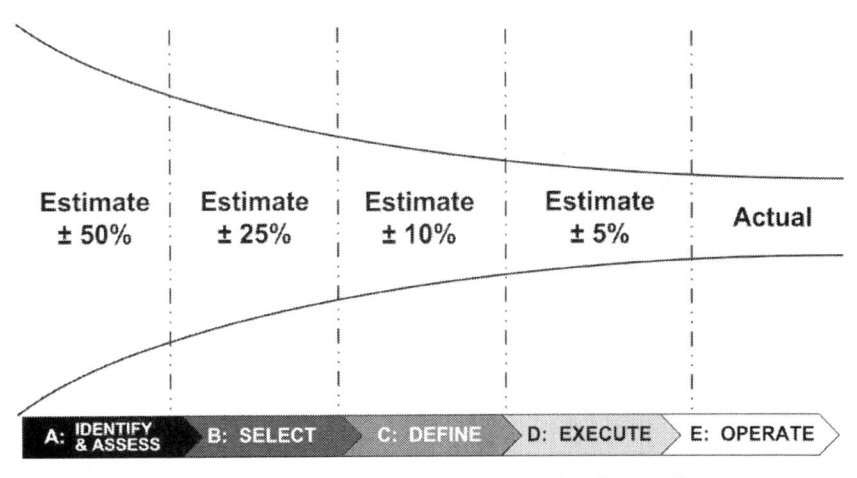

Figure 13 - Cost Estimate Maturity 'Funnel'

As the project reaches maturity, the Cost Estimates can be continually revised in the light of newer and more detailed and accurate information. As illustrated in the diagram, the accuracy of these estimates is usually very approximate at the outset, becoming fairly accurate by FID and as accurate as possible by the end of the EXECUTE phase.

Refinement usually begins as the various development concepts under consideration are worked, to arrive at a 'Preferred Concept'. At this stage, any obvious Operational Cost Issues should be identified and further refinement to the OPEX Estimate can be made.

In the later stages of the project, as Front End Engineering & Design (FEED) progresses, techniques such as Activity Based Cost Modelling (ABCM) are used to further refine the OPEX estimates. By the time the project reaches the FID, a fairly robust OPEX Estimate should be available.

The OR&A Team is also responsible for ensuring a suitable methodology for Operating Cost Management is in place prior to the introduction of hydrocarbons.

Consideration should also be given to 'hidden' costs of operating the asset such as taxation, tolls and levies.

Such costs could include production & transmission taxes, transaction based taxes, environmental charges & levies (such as the European CO_2 tax).

They may also include personal income taxes, which in some cases can drive salaries to an unnaturally high level, especially for expatriate staff.

OR&A Costs

The cost of implementing OR&A on the project is primarily the cost of manpower and the support costs incurred in providing that manpower. This is likely to include some or all of the following costs:

- Salaries,
- Office accommodation and services,
- Transportation (Office - Site - Vendor/Supplier - Rotation Leave),
- Residential accommodation costs

Obviously this includes the costs for (rotational and inter-site) and accommodation (dependent upon the location of the design team, vendors, suppliers and the construction team).

The OR&A Plan identifies the necessary OR&A activities which are incorporated into the Integrated Project Plan (IPP).

By allocation of the resources required to implement the OR&A activities and a reconciliation of project support costs, an accurate estimate of OR&A costs for each project phase can be determined.

Plant Costs - Lowest Quartile

At some point, because projects are always evaluated on cost and schedule, the asset will be benchmarked against similar competitors and projects will always try to achieve the lowest cost (CAPEX), unless there are overwhelming reasons for increased expenditure.

This is not always good from an OR&A perspective as the temptation will always be there to 'cut corners', often to the detriment of operability or maintainability. The normal benchmark parameters used are:

- $/bbl recoverable for facilities and drilling
- $/boed production rate for facilities CAPEX
- $/te topsides equipment, materials, fabrication (offshore projects)
- $/boed production for equipment and materials
- Ratio of equipment to final cost (onshore projects)
- Ratio of project management to total cost

Obviously, although asset output is mentioned, this will be the design performance figure, but longer term performance is not considered as there are no records to use.

At a later date, the benchmarks that are not considered here such as Availability (down-time), Reliability (long term performance) and sustainability will be evaluated, but this will be too late to affect the decisions made during the project.

Operations Involvement / Requirements

The involvement of the Operations Team in identifying specific Operations Requirements, concerns and challenges is essential to ensure that every aspect of each concept is considered and that the best concept is selected for further development.

It is the responsibility of the OR&A Team to identify the effects of these requirements, concerns and challenges on the Operability, Maintainability and the Life Cycle Costs of the asset.

Failure to carry out this 'due process' risks selection of a less than optimum concept and reduction in the viability and profitability of the completed asset. Typical Deliverables/Outputs for this Subject should include:

- OPEX Estimate (refined at each stage of the project),
- OR&A Cost Estimate (refined at the start of each phase),
- Contribution to selection of Preferred Concept on cost basis
- Identification of prohibitive cost issues for each concept under consideration.

Total Cost of Ownership (TCO)

Introduction

Design & Procurement decisions are auditable and are based on the principles of Total Cost of Ownership as detailed in the related British Standard and ISO Standard BS/ISO 15663 which were written in the form of a guide to the application of TCO principles.

Primary Objective of TCO

The primary objective of applying Total Cost of Ownership (TCO) principles is to evaluate and/or optimise the facility's life cycle costs whilst continuing to satisfy the specified project requirements, considering the three main cost elements:

- Capital Costs,
- Operating and Disposal Costs,
- Impact on Revenue Generation,
  (e.g. due to poor selection of equipment or equipment failure)

The aim is to enable objective decision making between options in all phases of an asset's life cycle especially in the early project phases.

Cost, benefit and risk assessment data will be necessary to review these options so informed decisions can be made. It is not a "free for all" exercise.

All selected options need to fulfil economic criteria and inhibit 'creep' in the project scope and an iterative process of re-visiting the business objectives is needed to preserve the overall life-cycle profitability of the project.

Scope of TCO

Standardisation of the approach and calculation methods necessary to establish and select systems and equipment that will give the lowest TCO over the life of the facility is achieved by optimising and selecting equipment design in relation to Capital Cost, Operating and Disposal Costs, and Impact on Revenue Generation (including cost of lost production).

These are defined in the TCO Standard BS 15663 Petroleum & Natural Gas Industries - Life Cycle Costing (parts 1, 2 & 3)

Application of TCO

The aim of applying TCO principles is to achieve and sustain Top Quartile performance.

To enable business organisations to maintain a competitive edge and to maximize their long term profitability, it is essential to provide facilities that meet all design requirements from the start that can be operated with the lowest possible OPEX, high availability and operability leading to the highest return on the initial investment.

The right foundations must be laid in the early project phases if this is to be achieved and where the TCO concept provides valuable support in the decision making process. TCO drives a structured assessment of project CAPEX savings against OPERATE phase OPEX costs and benefits.

Applying the TCO concept means evaluating alternative options and performing trade-off studies in the various development phases of a project in an effort to reduce operating costs (OPEX) but achieving the lowest life cycle cost could mean an increase in CAPEX.

Conversely, if the Project has an over-riding caveat to achieve absolute minimum CAPEX levels (whilst still resulting in technically acceptable equipment), in may have to accept higher operating costs (OPEX).

If this is the case the examination of the selected development options using the TCO approach will make the downside of the restrictions, from CAPEX based decisions, transparent. It will also identify additional critical activities required to maintain integrity and availability of the asset in the long term.

If TCO principles are used, accountability for the decisions made rests with the project manager (and later the asset manager). Anyone having a responsibility for TCO within an organisation should be made aware of the responsibilities and accountabilities assigned to them.

It is usual to assign the role of TCO co-ordinator as a project role rather than a named individual. Depending on the project phase, the individual may change but the role should remain. The strategy and policy for applying TCO should be included in the Project Strategy & Plan and should be defined (together with the means of evaluating whether the TCO objectives are being achieved.

6.1.3 Quality Management (1.03)

Introduction to Quality Management

The intent of Quality Management in a project is to apply the quality policies, objectives and responsibilities of the Project Owner to the project activities throughout the duration of the project.

The involvement of OR&A in project Quality Management is limited to ensuring that all OR&A activities (and activities monitored by the OR&A process) comply with the requirements of the Quality Management System, processes and procedures.

Quality Management processes should be applied equally to the execution the project and to the project outcome (the asset to be delivered).

It includes the processes and activities required to achieve the following purposes:

- **Quality Planning**,
  Identifying the quality requirements of the project and how the project will demonstrate compliance.

- **Quality Assurance**,
  Controlling and measuring and auditing quality performance against quality standards and operations requirements.

- **Quality Control**,
  Monitoring and recording the outcome of quality activities to assess performance and identify necessary remedial action.

The Quality Management process must provide a framework to manage the activities that ensure the project team consistently meets the project objectives and satisfies the promises made at every stage of the project.

The Quality Management System applied by the Project should conform to the principles and requirements of the ISO 9000 standard. The QMS should also promote the implementation of applicable International Standards, local regulatory requirements, company philosophies, engineering practices, and requirements.

The OR&A process is inherently interested in ensuring that a robust Quality Management System (QMS) in employed by the Project Management Team, not least because a failure to meet project quality requirements can have serious negative consequences for the asset and the project stakeholders.

Quality Philosophy

Recent Quality Management techniques are complementary to project management and both recognise the importance of the following tenets:

- **Quality by Prevention**
  Quality must be planned, designed and built into the project activities from the outset. Inspection plays a vital assurance role, but there is always a cost for corrective action.

- **Quality Improvement**
  There are a number of improvement models/initiatives (such as TQM, Six Sigma etc.), but all follow the basic structured methodology of the PDCA cycle (Plan, Do, Check, Act) process as pioneered by Shewhart and Deming.

- **Management Commitment**
  The success of any quality initiative, plan or process is dependent upon the demonstrated commitment and participation of the project management team.

- **Asset Owner Satisfaction**
  Achieving a successful handover to the asset operations team will rely upon the project being able to deliver an asset which satisfies the Operations Requirements and the promises made at FID. The QMS is instrumental in ensuring compliance and fitness for use to meet the requirements of the Project Owner.

Project Quality Plan (PQP)

The Project Quality Plan defines the quality policy, objectives and strategies for the project and though most of this plan should be taken from the Corporate Quality Plan and processes, it may have to be tailored to fit the project location, to address local legislation, rules etc.

The PQP provides information which is also pertinent to the Project Execution Plan and the Field Development Plan.

The PQP should describe how the project management team will implement the quality policy of the project, it should contain clear definitions of responsibilities, requirements for engineering reviews and preparation of the documentation necessary to ensure cost effectiveness, integrity and quality of facility design during project development (Front End Loading), detail design and subsequent phases.

Cost of Quality (COQ)

The term Cost of Quality (COQ) refers to the combined costs of the activities related to quality which extend over the lifetime of the project and the subsequent life-cycle of the completed asset.

Decisions made by the Project which fail to consider the long term effects of quality can have a dramatic adverse effect on the Operations costs (OPEX) due to early failure and breakdown of components and systems.

The cost to the project of implementing the quality requirements and thereby preventing the occurrence of failures versus the cost of correcting the failures which occur (especially after handover) is a clear advocate of the value of Quality Management and is summarised below:

COST OF CONFORMANCE	COST OF NON-CONFORMANCE
Prevention - building in quality: • Training of personnel • Documented processes • Correct equipment • Adequate time to do job right	Internal Failure - found by project: • Re-work • Waste (scrap) • Project milestones
Appraisal - assessing quality: • Non-Destructive Testing • Destructive Testing losses • Inspections	External Failure - found by asset: • Remedial Work (Warranty) • Liability for failures • Loss of reputation (business)
Costs to avoid failures	**Costs due to failures**

Clearly, OR&A is involved in all of the prevention mechanisms advocated by the Quality Management process as they affect other subjects within the OR&A sphere of influence.

OR&A is also concerned with the appraisal mechanisms, especially where they may affect Asset Integrity and Process Safety.

Less obvious is the OR&A involvement in the need to meet project milestones (to ensure the promises made at FID), the effect of failures after handover, the down-time due to the need for remedial work, repairs etc. and the loss of reputation to the Asset Owner in respect of the effects of those failures on stakeholders, neighbours etc. (e.g. flaring, spillages, emissions etc.).

6.1.4 Licences, Permits & Permissions Management (1.04)

The OR&A process in concerned with identifying, listing, obtaining, controlling and maintaining the necessary Licences, Permits & Permissions required for the entire project and asset life-cycle.

For every new project it is necessary to create a detailed list of all (Current) applicable regulations for the location (host country) of the opportunity.

These Acts, Laws and Regulations must then be examined to determine the extent to which they affect the opportunity and all of the Licences, Permits & Permissions required can then be identified and included in a Licences, Permits & Permissions Plan.

The relevant Acts, Laws and Regulations must include:

- Normative legislation, rules, regulations & standards
- Planning & development legislation, rules, regulations & standards (including access and Right of Way requirements)
- Human Resources legislation (occupational & labour laws),
- Financial Management (accounting rules and regulations, taxation and corporate law),
- Health, Safety & Environmental (HSE) (environmental law, safety regulations, asset rules, regulations and standards),
- Community Relations and Sustainable Development legislation, rules, regulations & standards.
- Development, Production, Processing, Storage, Transportation and Export legislation, rules, regulations & standards (including the necessary Licences, Permits & Permissions).

These are vitally important to the success of the venture and therefore it is always best practice to seek out local expertise to assist with these issues. Lead time and duration of validity are always thorny issues where Licences, Permits & Permissions legislation is involved and this must be given the necessary attention.

The Operations Team may not be responsible or accountable for obtaining early phase Licences, Permits & Permissions (such as those for development or construction), but since they could affect subsequent applications, it is pertinent for these to be listed (with suitable annotations) in the same register.

OR&A/Operations Team Involvement & Requirements

Not only is it is necessary to ensure the project team is aware of all applicable Acts, Laws and Regulations affecting the project and project activities, but the project must also be aware of the potential lead time and resources needed to obtain the necessary Licences, Permits & Permissions for the OPERATE phase in a timely manner and be aware of their duration/renewal cycles. Appropriate resources must be made available to ensure these issues can be dealt with efficiently.

Typical Concerns/Issues

Failure to identify applicable Acts, Laws and Regulations could result in unspecified transgressions and subsequent unforeseen penalties (financial, reputational and otherwise).

Failure to obtain the necessary Licences, Permits & Permissions in a timely manner could add significant delays and additional costs to the project.

Either of these constitutes a risk to the project and adequate cross reference to this subject needs to be included in the Risk Management Plan and Risk Register.

Typical Deliverables/Outputs

- Licences, Permits & Permissions (LPP) Plan detailing the process for identifying LPP requirements and maintaining compliance.
- LPP Matrix/Register containing all of the pertinent details including acquisition cost, duration and renewal cycle.
- Rights of Way/Access (to be included in Matrix/Register).
- Secure repository and monitoring system for LPP.
- Responsibility/accountability assigned to appropriate Technical Authority from Asset Organisation.
- The detailed list of all (Current) applicable Acts, Laws and Regulations for the location (host country) should be reviewed, updated and reissued annually.
- Confirmation of LPP compliance should be verified by regular reviews of the plan and status and documented in a formal Compliance Report.

6.1.5 Flawless Project Delivery (1.05)

Flawless Project Delivery (FPD) is a key element of the OR&A process.

The aim of Flawless Project Delivery (FPD) is essentially to ensure a smooth, first time start-up of the asset, followed by a steady ramp-up to sustained optimum performance at the design intent.

FPD is a key enabling process for successful project delivery. It is intended to ensure a structured and timely execution of activities that should lead to a 'right first time' start-up and subsequent stable and sustainable operations, and maintenance of the completed asset.

Clear commitment of Senior Management from the earliest stage of the project development process is essential for FPD because of the potential contractual implications.

The FPD process identifies and assesses risks to Flawless Project Delivery and to implement appropriate mitigation measures.

The 'business case' for an FPD is best illustrated by the graph shown below

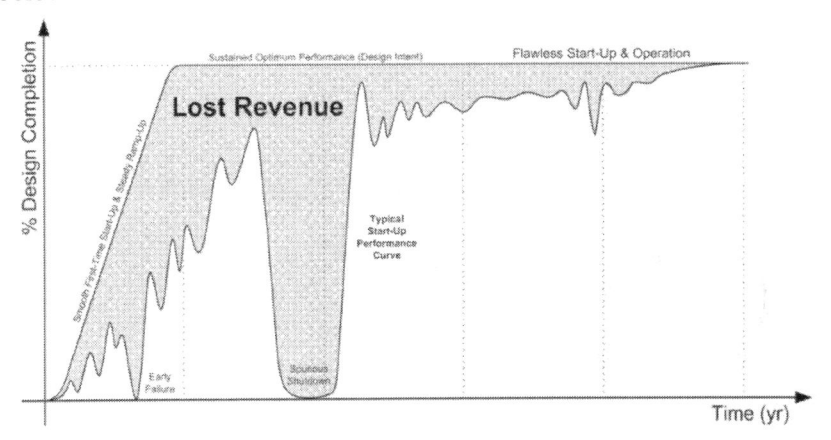

Figure 14 -   Flawless Start-Up

Flawless Start-up and Operation is represented by the upper line on the graph, an example of a smooth start-up and operation of facilities to the design intent, usually presented as a part of the business case a project.

However, in reality, the performance is often typically represented by the lower line, assuming a Time axis graduated in months.

The graph also illustrates the loss of production/revenue (the shaded area between the two lines) which can often be evaluated in the order of tens and sometimes hundreds of millions of pounds.

Additionally, a risk to the reputation of the asset owner and potential HSE issues can compound the financial losses.

How FPD is Implemented
Essentially, the root causes of the failure of many projects to achieve a Flawless Start-Up originate from decisions and activities in the early (pre-FID) development phases. If these are not resolved in the early phases, the result will be delays, additional costs and a failure to achieve FPD.

The remedy for these root causes (often called Flaws) requires recognition of these issues and mitigating action to be taken to tackle the causes before they affect project delivery.

Many of these causes have already been recognised and can be prevented by implementation of key Lessons Learned (from previous projects). Others can be resolved by emphasizing specific techniques to avoid common errors and omissions.

Typical root causes or Flaws are related to a number of key performance areas, such as those listed below:

- Tightness;
- Cleanliness;
- Integrity;
- Operability;
- Health, Safety and Environment (HSE);
- New Technologies (Prototypes);
- Complexity;
- Testing;
- Experience;
- Coinciding events;
- Information;
- Integration.

First and foremost, for those root causes (Flaws) identified as applicable to the project, an assessment of their criticality in respect of impact on the project cost and schedule must be made. This allows the focus (and the mitigating action) to be placed on those causes which will be most cost effective in relation to the cost of that mitigating action.

6.2 Operations Management (2.0)
6.2.1 Operations Management & Planning (2.01)
Operations Management

The objective of the Operations Management component of OR&A is concerned with the ensuring that preparations for the management of production activities in the OPERATE phase have been put in place ready for the introduction of hydrocarbons (except for fuels and lubricating oils) into the asset.

To complete this objective is essential to fully understand, at an early stage of the project, the operating environment in which the project will be developed as this will dictate what operations can be undertaken and how they can be carried out.

Such a detailed understanding of the operating environment must include an evaluation of factors, (including Regulatory compliance), which may affect future operational activities and operational performance.

Operational drivers alone are not the single factor in concept selection but can have a significant impact on the overall life cycle value and must be carefully documented in the Operations Philosophy and addressed when selecting a preferred concept for further development.

Other aspects of the operating environment which have a bearing on concept selection may include commercial arrangements, production agreements, characteristics of the produced fluids, order of magnitude of the development and a detailed assessment of local resources.

As the project progresses, the potential impact of the project must be assessed against key operational drivers (i.e. Availability, Integrity, Operating Expenditure (OPEX) and HSE issues) to determine if a sound basis for the decision to move the project into the SELECT phase exists.

The main activities of the OR&A team include creating, developing and implementing the necessary plans, processes and procedures for the OPERATE phase. It is also essential that they ensure the Operations Team are fully prepared to manage and operate the asset responsibly and in a safe, sustainable and environmentally friendly manner.

The OR&A team must be integrated into the project team and ensure they are both visible to, and active within, that project team at all times. Their participation in the key project activities will ensure that the Operations Requirements are included from the outset.

The OR&A Plan provides a baseline of necessary activities which is subsequently incorporated into the Integrated Project Plan to ensure it is fully aligned with the project activities, timeline and milestones. It includes all Operations related activities in subjects which have a bearing on that safe, sustained, environmentally responsible and sustainable production.

Integrated Planning

Integrated Planning, (also known as Integrated Activity Planning), specifies the activities required to prepare the custodians (owner/operator) of the asset under construction (and the supporting organisation) such that, at the point of delivery/handover, that organisation is fully prepared.

By default, this includes participation in the Project activities to ensure the needs of the Asset Operations Team (to enable them to do this) are recognised and addressed during the development of the asset.

The objectives of Integrated Planning are summarised below:

- Optimisation of deployment of operations resources.
- To maximise the availability of the production facilities.
- To control and reduce costs.
- To provide a method for delivering accurate production forecasts.
- To reduce the risk exposure of the operations team.
- To provide a common reference for controlling operations activities.
- To provide a means of comparing and analysing the plans and schedules for the OPERATE phase.

The Integrated Planning process includes all activities that impact field operations and/or make use of shared critical resources, (e.g. combining well engineering and well services, maintenance and inspection of facilities including pipelines, engineering (both major and minor projects) and exploration).

In the early phases of the project, (IDENTIFY, ASSESS & SELECT) typical outputs or deliverables from this subject should include:

- (Preliminary) Operations Assessment
- (Preliminary) Operations Strategy & Plan
- (Preliminary) Operations Philosophy
- A contribution to preliminary Project Execution options
- Operations Requirements (refined as the project progresses)
- Concept Selection Criteria
- Contribution to the Field Development Plan (FDP)

In the DEFINE phase, during FEED, these documents will be further refined and the typical outputs or deliverables will include:

- Contributions to the Project Execution Plan (PEP)
- Contributions to the Basis for Design (BfD)
- Contributions to the Invitation to Tender (ITT)

In the EXECUTE phases, many of these documents will be further refined and the typical outputs or deliverables will include:

- Development of the Asset Reference Plan (ARP)
- Development of the Operations Management System (OMS)
- Development of the Hydrocarbon Accounting System
- Identification of other IM/IT systems for operations

Finally, population of the necessary operations systems with asset data must also be completed before the introduction of hydrocarbons (excepting fuels and lubricating oils) to ensure the baseline data and early operating history of the asset equipment is captured.

Operations Philosophy

The Operations Philosophy is a high level overview which describes how the Operations Team intend the wells and facilities to be operated.

The Operations Philosophy includes references to standards, policies, procedures and attitudes that affect the operations and maintenance of the asset, wells and facilities.

The general operating and facilities maintenance philosophy and policies defined in the Operations Philosophy will include details to be included in documents such as the manning philosophy, (e.g. for a development involving near shore platforms and might make recommendations such as a preference for day travel rather than rotational platform accommodation).

The Operations Philosophy also briefly describes the proposed infrastructure and the current plans for development. The document should be revised in every phase of the project, both to ensure the project is following the intended philosophy and to record any changes to the policy and the proposed development.

6.2.2 Production Operations (2.02)

In the early phases of the project, (IDENTIFY & ASSESS), the involvement of the Operations Team in the project will be minimal as the team is likely to be just the Operations Readiness Manager/Advisor and even then, this could be only a part-time assignment.

At this stage, a number of possible concepts will be under consideration as part of the Opportunity Framing Process. An evaluation of each opportunity and the constraints (from an Operations perspective) need to be identified and provisional Block Flow Diagrams (BFD) for each concept created in preparation for concept selection activities.

During this process Operations experience based knowledge is a vital input must be available during the conceptual design phase of a new development.

In the SELECT phase, for each concept under consideration, an evaluation of the operability and maintainability will be required. This will produce specific Operations Criteria for each concept to assist with selection of a preferred concept for further development.

For the selected concept, the Operations Criteria will subsequently be refined to form the basis of the specification of the Operations Requirements to the design team and will include the following:

- Site location and preliminary location of major equipment,
- Development of the Block Flow Diagrams,
- Development of the Process Flow Diagrams,
- Development of the Heat and Material Balance documents,
- Confirmation that the preferred concept satisfies the Operations Requirements and Selection Criteria.

At this point, the Selection Criteria will address the steady state operations of the facilities by defining the following parameters:

- The operations objectives, strategies and work processes,
- Manning strategy (rotations/work cycles/local skills, etc.),
- Operability and Maintainability of the proposed facilities
- Key operational risks and HSSE hazards
- Estimated OPEX

Much of this information will be recorded in the preliminary Operations Philosophy and will usually form the basis for the Operations Management System.

In the DEFINE phase, after the preferred concept becomes a project, the detail of the project and the size of the project team increases proportionally.

The OR&A Team will be supplemented to ensure the Operations Requirements are recognised and incorporated into the design.

These requirements should begin with a clear and concise definition of the operating and maintenance requirements with the aim of ensuring that the asset is accessible, operable and maintainable regardless of its location.

The Operations Requirements will prescribe commonality in standards, codes, policies, procedures and practices across the project, both for economy of operation and to reduce the burden upon the operations staff.

During this early design phase the Operations Reference Plan should be created and should address the entire Integrated Production System. It will provide strategy and policy input to project planning, well and facility design and equipment selection.

Cost estimates for operating options provide the project with an opportunity to optimise opportunities. It is a high-level route map for the life cycle of the integrated production system. It also provides guidance on Operations manpower and skills requirements.

At the end of the DEFINE phase, the project will have to pass through a decision process to determine whether it is a viable proposition and to obtain authorisation to commit the financial resources required to complete the project. This is usually known as FID, the Financial Investment Decision.

Beyond this point, any changes or modifications will be more expensive and difficult to make due to the volume of work done and the likelihood that those changes will mean changes to equipment that is being or has been constructed/ordered.

As the project moves into the EXECUTE phase and detailed design begins in earnest, the volume of the work for the entire project increases significantly. At this point, the OR&A Team will usually increase proportionately, but will also include a number of Asset Operations Staff.

At this point, direct input from the Operations and Maintenance team is vital. Selection of components/equipment will determine the operating and maintenance activities, and the knowledge and skills required to carry out those activities.

The project documentation will need to be reviewed and revised to ensure the latest information from the project and will then provide the general guidelines for the Design team, however, experienced Operations and Maintenance personnel should also be involved directly in the detailed design process.

During the EXECUTE phase, most of the commissioning, operating and maintenance documentation will be prepared, based on the equipment specified by the design team. It is essential that the procurement process includes requests for vendor information to cover the following aspects:

- Handling & Preservation,
- Installation,
- Pre-commissioning,
- Commissioning,
- Start-Up,
- Operations,
- Troubleshooting,
- Maintenance,
- Spare Parts,

This information should be available before the equipment arrives at site (particularly the Handling & Preservation and Installation manuals).

The preliminary Asset Reference Plan should also be prepared during the early EXECUTE phase and updated continuously throughout the phase.

The ARP makes statements on well and facility systems, contracting and manning policy and defines the strategies, plans, policies and guidelines and includes a more robust life cycle cost projection.

All of these activities require experienced staff assigned to the Detailed Design, Construction and Commission and Handover activities.

Ideally, the same staff should follow through the project from detailed design to operation of the asset. This will ensure the transfer of experience and also ensure that quality production operations and maintenance data is collated, analysed and fed into the various stages of development.

Life Cycle Considerations (Post Handover)

By the time the EXECUTE phase commences and Detail Design begins in earnest, the Operations Management emphasis will be on ensuring that the necessary processes, procedures and systems are defined and in place to support operations activities from the point of start-up of the asset. This should include the following:

- Well Operations and Maintenance
    o Well W/O Execution
    o Subsurface Well Testing
    o Facilities Operation & Inspection
    o Preventive Maintenance
    o Corrective Maintenance
- Asset Management Excellence
    o Oil & Gas Production Management
    o Oil Field Water Management
    o Gas Storage Management
    o Weekly Production Reporting
    o Monthly Reporting (Costs & Production)
- Asset Development Excellence
    o Well Deliverability Improvement
    o Facilities Efficiency Improvement
    o Total System Performance Improvement
    o (Subsurface) Production Optimization
    o Re-Development Management
- Planning Excellence
    o Facilities Shutdown Management
    o Well W/O Planning
    o W/O Scheduling
    o Production and Cost Forecasting (MTP)

6.2.3 Production Chemistry & Flow Assurance (2.03)

Introduction to Production Chemistry

Production problems occur in every oil and gas facility and the magnitude of these problems varies according to the nature and composition of the process fluids; oil, associated water and gas and the type of production operations being carried out.

The problems also vary with time, in response to changes in temperature and pressure and in response to changes to the amount and composition of associated brine.

Knowledge of the composition of the fluids and the prevailing environmental conditions will provide input to predictive models that can help to define the issues and provide appropriate recommendations for action to minimise the damage. Generally the production and flow assurance problems experienced fall into four categories:

- **Physical (Emulsion & Foam)**
  Based on the physical properties of the process fluids. These are processing issues of which foam, emulsion and flow viscosity are typical examples. Normally, chemical additives are used in measured amounts to help to overcome the effects of these types of issue.
- **Corrosion / Structural Integrity**
  Typically, corrosion affects the structural integrity of the plant (and the safety of the workforce). These issues are typically addressed by the use of measured dosing of corrosion inhibitor at key locations in the process.
- **Fouling / Flow Assurance**
  Fouling is defined as the deposition of any unwanted matter in a system, including scale, corrosion products, paraffin wax, asphaltene, biofouling and gas hydrates.
  Chemical inhibitors are added to reduce the rate of fouling and solvents are added to remove existing deposits.
- **Environmental / Economic**
  All discharges have environmental and economic consequences and must be carefully monitored, controlled and treated. Typically, these include oily water discharges or the presence of sulphur compounds such as hydrogen sulphide and mercaptans in vented gasses.

It essential that Production Chemistry issues are considered at every stage of the project to ensure that problems such as corrosion, flow assurance or increased costs due to the need for excessive volumes of production additives (production chemicals) are not inadvertently built into the design.

The practicalities of Production Chemistry include:

- Ensuring the production fluid composition is fully analysed, from the outset, to identify possible Production Chemistry requirements and flow assurance issues (wax formation etc.),
- Cost estimates, based on the analysis of the process fluids and projections of possible Production Chemistry requirements are prepared and updated at each stage,
- Confirming the design team recognises the importance of Production Chemistry to the long term operation of the asset,
- Confirming the design team is aware of and catering for the Production Chemistry requirements,
- Inclusion of specific hardware in the design of the production facilities (e.g. desalting, sulphur recovery, sour trim equipment etc.).
- Provision for flow assurance operations such as pigging,
- Provision of sample points, (at key locations) including accessibility to those sample points,
- Provision for the injection of chemicals and other additives at key points in the process,
- Facilities to aid chemical injection, such as injection points, chemical dosing pumps,

The OR&A Team (on larger projects) will usually include an engineer with Production Chemistry expertise, but even on smaller projects this should not be overlooked.

The following pages contain a little more detail of the typical reservoir fluids types of materials/issues that need to be dealt with, either by design or treatment, but it should be remembered that the cost of these can be significant to the Asset Operator over the lifetime of the asset.

Typical Reservoir Fluids

Hydrocarbon Fractions

The hydrocarbon phase of reservoir fluids consists of the n-paraffin series, which is divided into 3 groups (gasses, liquids and solids) as listed below:

C1 (methane) CH_4 C2 (ethane) C_2H_6 C3 (propane) C_3H_8 C4 (butane) C_4H_{10}		gas	Gaseous compounds from the lower end of the alkane or paraffin homologous series of hydrocarbons, (members of a homologous series differ from each other by one carbon atom and two hydrogen atoms)
C5 (pentane) C_5H_{12} C6 (hexane) C_6H_{14} C7 (heptane) C_7H_{16} C8 (octane) C_8H_{18}			Liquid fuel fractions, volatile and very flammable (generically referred to as Naphtha). Petroleum and gasoline are terms also used to describe these fractions.
C9 (nonane) C_9H_{20} C10 (decane) $C_{10}H_{22}$ C11 (undecane) $C_{11}H_{24}$ C12 (dodecane) $C_{12}H_{26}$ C13 (tridecane) $C_{13}H_{28}$ C14 (tetradecane) $C_{14}H_{30}$ C15 (pentadecane) $C_{15}H_{32}$		liquid	Higher viscosity liquid fractions less effective as a motor fuel but safer to handle due to lower flammability. These include paraffin, aviation and diesel fuels. Flow problems occur at lower temperatures due to high melting point of the liquids (-53°C to +10°C).
C16 (hexadecane) $C_{16}H_{34}$ C17 (heptadecane) $C_{17}H_{36}$			Fractions described as diesel, gas oil and fuel oil often used in central heating systems and marine fuels. Flow problems can occur at lower temperatures due to high melting point of the higher alkane liquids (+18°C to +21°C).
C18 (octadecane) $C_{18}H_{38}$ C19 (nonadecane) $C_{19}H_{40}$ C20 (icosane) $C_{20}H_{42}$		solid	Lubricating oils and paraffin waxes, which are more viscous, compared to fuel oils. The higher fractions are bitumens, which are solids at room temperature and are waterproof. Bitumens are often used on roads and roofs.

Typical Reservoir Fluids (continued)

Condensate
The term 'condensate' relates to the liquid hydrocarbons that condense to the liquid state when the natural gas is produced from the reservoir and the temperature of the produced gas falls below the hydrocarbon dew point temperature of the gas.

Crude Oil
Crude oil is a naturally occurring, flammable liquid consisting of a complex mixture of hydrocarbons of various molecular weights and other liquid organic compounds, found in geologic formations beneath the Earth's surface.

Crude oil that contains no water is called 'dry crude'. It is unusual for a field to produce dry crude for an extended period.

Inevitably water, in the form of natural formation brine will be produced, usually as a water-in-oil emulsion or as free, uncombined brine.

Produced Water
Natural formation brines (also known as connate waters) contain dissolved salts and dissolved gases.

The common dissolved salts are sodium, potassium, calcium and magnesium chlorides, sodium bicarbonate, sodium sulphate and lesser concentrations of strontium, barium and/or iron chlorides.

The content of dissolved salts can vary from about 1000 mg/L to sodium chloride saturation at near 300,000 mg/l.

Carbon dioxide and hydrogen sulphide are common dissolved gasses, however Oxygen is absent from deep subterranean formations but may be present in the topside facilities due to ingress of air from the atmosphere.

Bacterial populations will generally be present in injected brine although in extremely small proportions.

Silt and clays together with scale minerals, iron corrosion product, iron sulphide and precipitated asphaltenes can also be present in produced fluids.

Typical Reservoir Fluids (continued)

Sand

Natural formation fluids can also contain varying quantities of sand entrained in the production fluids.

Although produced sand is inert, build-ups will occur where any form of restriction is present and this can cause flow restrictions and blockages. The flow of wells must be controlled to reduce sand production wherever possible.

Sand can also be abrasive in nature and can cause erosion in parts of the production system where the velocity of the produced fluids is high enough.

Flow Assurance plan(s) related to wells should therefore include a Sand Management Strategy & Plan.

Emulsions

Due to typical oil & gas reservoir formations, as crude oil is produced from a reservoir it tends to become mixed with either natural formation water and/or injection water and this mixture is called an emulsion.

The purpose of the process operations is to produce 'dry' crude oil for export (and to dispose of produced water by the most appropriate method) to maximise the value of the product and reduce costs.

The efficiency with which an emulsion can be broken down is dependent on many factors including:

- The physical and chemical properties of the crude oil;
- The temperature of the production fluids;
- The distance between the reservoir and process facilities;
- The degree of turbulent flow experienced between the reservoir and process facilities;
- The presence of solids (sand, clay, bacteria, scale, asphaltenes, corrosion product, naphthenates) and / or natural surfactants which act to stabilise the emulsion.

The separation process is the primary method of separating oil and water (and gas), but to further remove the impurities from the crude oil component or from the separated water, more detailed processes may be required before export of the product and disposal (or re-injection) of the water can be considered.

This may include processes such as desalting, dehydration, demulsification or de-oiling to achieve the crude oil product and residual produced water specifications.

Foam

Naturally occurring dissolved gases in the reservoir fluids behaves as a single liquid phase and is in equilibrium. When the crude is produced, agitation as it flows from the reservoir to the first stage separator causes the dispersion of the gas into the crude oil, which results in the formation of foam. Problems associated with excessive foaming can include:

- Reduction of separator capacity;
- Carry-over of oil;
- Fouling of gas flares;
- Increased costs of operation;
- Pump cavitation.

Corrosion

Corrosion is a chemical or electrochemical reaction on a metal by something in the immediate environment and is one of the most serious issues affecting asset integrity.

A failure of asset integrity due to corrosion, resulting in a release of hydrocarbons into the environment could result in serious harm to personnel, plant & equipment and the environment.

This would inevitably lead to damage to the reputation of the Company, loss of revenue and the cost of repair or replacement of equipment in a potentially hostile and inaccessible environment.

Electrochemical corrosion occurs at the solid/fluid interface in water, water/oil and gas systems. It can occur in H_2S (sour) systems, in CO_2 (sweet) systems, or in a combination of both.

The consequences of electrochemical corrosion can be severe and include:

- General metal wastage (general corrosion),
- Pitting,
- Embrittlement of steel and surface cracking.
- Equipment failure.

The main drivers that determine corrosion rates of steel in oilfield water systems are:

- Acidity/alkalinity (pH),
- Temperature,
- Hydrodynamic flow conditions,
- Presence of H_2S,
- Traces of oxygen.

The concentration of Carbon Dioxide (CO_2) in any associated gas is critically important, as is the pressure within the system.

These two factors determine the partial pressure of CO_2, which in turn determines the CO_2 concentration in solution and hence the pH of the associated aqueous phase.

Carbon Dioxide (CO_2) induced corrosion dominates the general corrosion mechanism in most oil & gas systems, can be adequately modelled if significant input parameters are known.

However, corrosion initiated locally by bacterial activity and/or Hydrogen Sulphide (H_2S) is more difficult to quantify.

Fouling & Flow Assurance

The term 'scale' is generically used to describe unwanted material from the produced fluids that is deposited on any surface within the production system.

The products of corrosion, organic materials and common limescale could all be described as scale. However, it is more usual in the oil & gas industry for the description to be limited to materials formed from inorganic compounds, or more specifically 'mineral scale'.

The deposition of scale and sludge in a reservoir or in oil production systems is a common, widespread and potentially serious problem and under certain conditions, produced water can also lead to the formation of scale in the production system. Carbonate scales, most commonly calcium carbonate, can result from temperature and pressure changes in the natural formation water as it passes through the process system.

The problems caused tend to be regarded as flow assurance issues, as listed below:

- Reduces cross sectional area of the pipe;
- Increased friction between fluids and internal surfaces;
- Restricted fluid flows due to exchanger tube obstructions;
- Reduced heat transfer on heat exchanger surfaces;

All of these issues need to be addressed as early as possible in the design process to ensure the appropriate design decisions are taken. The operations team must be involved in such discussions to ensure the long term operability of the asset.

Predictions from software modelling can be valuable to identify whether the produced water has a propensity to form scale at any point in the production process from well-bore to topsides conditions.

It should be remembered that the magnitude of the predicted scale formation can vary enormously for a number of geographic, climatic and other natural conditions. The operation of the production facilities can also either encourage or discourage the formation of scale.

Typical Scales	
Calcium Carbonate	
Where formation water is naturally very rich in calcium, mixing of injection water and formation water can promote the deposition of calcium carbonate scale. When the pressure of the produced water is reduced, dissolved CO_2 in that water is released. The subsequent increase in pH, rather than the direct effect of the pressure reduction, that causes calcium carbonate (calcite and/or aragonite) to form at relatively low pressures. Where increases in fluid temperature (heating) and depressurization occur simultaneously, (such as in the crude oil stabilization process), this effect is significantly increased.	Typical mitigation strategies include the application of scale inhibitor or dissolving the scale by acid treatment.
Sulphate Scale	
The formation of sulphate scale typically occurs when sulphate-rich water is re-injected (to maintain reservoir pressure) and is prevalent in the production risers (tubulars). Formation water is often rich in barium, strontium and/or calcium but naturally contains very little sulphate; conversely seawater containing sediment or decaying organic material is sulphate-rich. Sulphate scale is insoluble in acid and pressure, temperature and pH have less effect on sulphate scales than on carbonate scales.	Scale inhibitor squeeze treatments, treatment with specialist dissolvers or, in severe cases, installation of desulphurisation technology are the only alternatives to expensive work-over to combat this type of scale.

Barium Sulphate	
Barium sulphate is highly insoluble and deposits will form at any location where temperature drops occur throughout the production processing plant. Usually barium and/or strontium scale formation usually results from mixing of incompatible waters. Barium sulphate can sometimes be radioactive because these scales co-deposit and greatly concentrate radium isotopes and their 'daughter' radio-nuclides	Scale inhibitor squeeze treatments, treatment with specialist dissolvers or, in severe cases, installation of desulphurisation technology are the only alternatives to expensive work-over to combat this type of scale.
Iron Sulphide	
Iron Sulphide scale is deposited where microbial enhanced corrosion has become a serious problem. Scale is a by-product of the reaction between iron oxide from corrosion and Hydrogen Sulphide (H_2S), a by-product of Sulphate Reducing Bacteria (SRB).	Treatment for iron sulphide is application of a specialist chelating and dissolution agent followed by microbial control with biocide.
Sodium Chloride (Halite)	
Sodium Chloride (Halite) scale is caused by cooling or evaporation of extremely saline formation brines.	Removed by dissolving and/or flushing with warm water.

Wax & Asphaltenes

Some produced crude oil contains 'heavy' or 'waxy' components (C_{18} to C_{40}) such as paraffins, waxes and bitumens which may result in:

- Deposition of these heavier fractions in the reservoir, tubulars pipelines and process facilities;
- Pumping difficulties caused by increased crude viscosity and solid particles.

The waxes present in most crude oils (typically n-Alkanes) are specific to each variety of crude oil (and each different reservoir). These have a significant impact on the deposition characteristics. Paraffin solubility not only depends on the composition of the crude but also the temperature and pressure.

The Wax Appearance Temperature (WAT) is the temperature at which wax solids are precipitated by the cooling crude oil. The deposition of wax leads to flow assurance problems in terms of reduced and impaired production.

The precipitation of other naturally-occurring hydrocarbon fractions of crude oil known as asphaltenes and resins is strongly dependant on the crude composition, pressure and temperature. The operating problems are similar to waxes but are more difficult to predict and to treat.

Asphaltenes differ between crude oils from different reservoirs, but retain common properties and general chemical structures regardless of crude origin.

In comparison to paraffin, asphaltenes are generally soluble in toluene but insoluble in lower n-alkanes such as pentane and hexane.

Microbiology

Frequently, oil production systems experience severe problems due to the growth and proliferation of bacterial populations.

Microbiology is the study of microscopic living organisms and broad terms, this means those organisms with a diameter smaller than 0.1 mm.

There are three important types of bacteria that occur in oil production systems and these are listed below with a brief description of the associated problems, their growth and proliferation:

Sulphate Reducing Bacteria (SRB)
Sulphate-Reducing Bacteria (SRB) produce hydrogen sulphide as the main product of their metabolism and this poses a significant health and safety risk to humans and can also result in the formation of metal sulphide sludge and scale deposits.

There are various types of SRB, which can proliferate over a range of different temperatures and salinities, though lower temperatures and lower salinities constitute the greatest risks.

Though SRB and other bacteria exist in 'planktonic' form, they also have a propensity to form surface slimes, termed biofilms, which can produce localised high concentrations of sulphides and other harmful products and it is these 'sessile' biofilms that give rise to Microbially Influenced Corrosion (MIC).

General Aerobic Bacteria (GAB)
Slimy exopolymer, ideal for biofilm formation because it enables bacteria to slick to surfaces in the form of biofilm can be produced in large volumes by General Aerobic Bacteria (GAB) and can use a wide range of carbon sources for growth and energy production.

Slime can lead to fouling and blocking of filters, lines, injection pores etc.

The formation of concentrated cells of oxygen can lead to the development of corrosion under the slime deposits and promote an ideal environment for Sulphate-Reducing Bacteria (SRB).

Since General Aerobic Bacteria (GAB) can also live and grow in oxygen-free conditions, where they may produce organic acids by fermentation reactions (i.e. converting one form of organic carbon into another to produce energy), the alternative name General Heterotrophic Bacteria (GHB) is more accurate and should be used.

Acid Producing Bacteria (APGHB)

A sub-set of the GAB/GHB are Acid Producing Bacteria (APGHB), which are also implicated in the corrosion process as they locally metabolically secrete organic acids, which can become trapped under bacterial biofilms and promote corrosion.

It is common for all three types of bacteria (SRB, GAB/GHB and APGHB) to grow simultaneously in biofilm and to exert an enhanced corrosive effect by their combined action.

Hydrogen Sulphide (H_2S)

The injection of seawater into a hydrocarbon-bearing reservoir (usually to increase production by increasing reservoir pressure) can lead to the growth of Sulphate-Reducing Bacteria (SRB).

These are anaerobic bacteria that chemically reduce sulphate ions to sulphides, with the naturally associated oxidation of organic carbon compounds.

Production of these sulphates in the reservoir leads to the appearance of hydrogen sulphide, sulphide ions, metal sulphides or a range of these, depending on the reservoir conditions.

Sulphide production in the reservoir leads to 'souring' of the produced gas and fluid. Reservoir souring can also occur naturally and from contaminated drilling mud during drilling operations.

Hydrogen Sulphide as a solution is a direct cathodic reactant in the corrosion process and high localised corrosion can occur in situations where a neutral pH and the absence of oxygen would otherwise be regarded as a benign environment.

Environmental Discharges

Usually any discharges of gas, effluent or other waste is regulated by law and will have severe consequences if those rules are ignored or even inadvertently transgressed.

All such discharges must be carefully monitored, controlled and treated to ensure compliance with those laws and rules. Loss of an operating licence would cripple any operating asset and could be a serious penalty for repeated offences.

Typically, strict limits are imposed on the most obvious of these discharges, such as oily water or the presence of sulphur compounds such as hydrogen sulphide and mercaptans in vented gasses.

6.2.4 Wells (2.04)

It many Exploration & Production companies, because oil reserves are a closely guarded secret, the reservoir and drilling departments are allowed to act in isolation. Whereas this can be good for security of the company assets, it may also be bad for projects and long term oil production performance (and therefore have a negative effect on profits).

Operating producing wells and optimisation of the reservoir and production rates of the wells may differ between concepts being evaluated. The ability to operate the wells according to the well design envelope may be constrained due to facility flow assurance considerations, facility bottlenecks and/or contractual (delivery) requirements etc.

Other limitations such as maximum production rates to avoid damage to wells (e.g. sand production) may be set out in the field depletion plan (or field drainage plan as it is sometimes referred to).

The project OR&A representative must contribute to early well design discussions, not only offering advice on well operability issues and early well concept designs but also being made aware of those limitations which may affect the advice given.

It is important that all of the relevant disciplines attend all reviews and workshops to ensure the operational functional requirements are aligned (as detailed in the project Operations Framework).

Well operability issues that may impact upon concept design must be evaluated in terms of risk to HSE, project delivery (quality and schedule) and cost. It is essential to achieve early identification of:

- Key operational drivers;
- Key operational risks;
- Key operational opportunities to improve well delivery.

Early requirements from an Operations perspective need to be clearly understood so that they can be included in the identification of well design concepts and help the project team to ensure the selected concept(s) are robust.

These include the identification of clear selection criteria (including operational requirements) the concept selection process and to ensure that they are included in the Operations Philosophy.

Typically, an early draft of the wells design is prepared by the wells discipline and an integrated view is difficult to form due to limited definition around each discipline.

This makes it even more important for the OR&A engineer to identify key operational issues that impact the well operability and ensure that these are carried forward from the SELECT phase into the project Basis for Design (BfD) documents.

Additionally, operability of wells utilising new technologies (e.g. with increased levels of data acquisition and control capability) must be reviewed in order to create awareness of the increased level of capability the production system concept may be required to handle, within the project and operations teams.

Studies to fully appraise any new technologies, modes or methods in the processing facilities should be started as early as practicable, since operability of the asset (resulting in increased/decreased cost/schedule of the development) may be significantly impacted by decisions made during this phase. These may include:

- Operating modes (manned or remotely operated);
- Well access (e.g. sub-sea vs. platform);
- SIMOPS (during construction and during future interventions);
- Well surveillance and intervention requirements;
- Required production availability as described in initial licence agreements discussions etc. (e.g. PSAs);
- New / novel well technologies;
- Logistics requirements and availability.

Typical Concerns/Issues

One of the main operational concerns in the early days of any project is being confident of the composition of the production fluids and flow rate/production data as it becomes available.

Failure to obtain thorough and detailed data on the production fluids prior to the SELECT phase can result in a great deal of rework if the design options do not fully recognise issues and properties in those fluids that need to be specifically addressed by the design.

The effect of such failures could be critical to the continued success of the project and the eventual viability/profitability of the asset.

Typical Deliverables/Outputs

Ideally, the well and reservoir management team will welcome the involvement of the OR&A Team as the representative of the Operations Team in the IDENTIFY & ASSESS phase.

Typical deliverables from this co-operation should include:

- A good working relationship is established between the well and reservoir management team and the OR&A Team.
- Operations Requirements for the long term operation of the wells is clearly defined.
- The Wells section of the preliminary FDP is reviewed and comments/contributions made as appropriate.
- The Flow Assurance Plan(s) related to the long term operation of the wells reviewed (e.g. Sand Management Strategy).
- Risks/opportunities around new well technology are identified, studied and fully captured in the Risk & Opportunity Management Plan for the project.
- Endorsement of the early concept well design by OR&A team.

6.2.5 Infield & Export Pipelines & Manifolds (2.05)

The purpose of this subject in respect of OR&A is to ensure that the design of the pipelines and manifolds for the proposed development are designed in such a way as to maximise the potential of the field and the export capabilities whilst minimising the initial cost (CAPEX) and the cost of operating and maintaining the system (OPEX).

In the first instance, it is essential for the Opportunity Framing Team to understand the operating environment of the proposed development. In the first instance, this will be based on the characteristics of the reservoir fluids and the magnitude of the opportunity.

The Opportunity Framing Team will then consider the necessary processes which must be addressed from the wellhead to the point of sale, in the form of a high level Block Flow Diagram for the proposed project.

From this point, the Opportunity Framing Team will consider the physical and geographical environment and the locations of the wells, processing facilities and export facilities and develop a number of options to be considered for the future development.

The options must then be refined in the light of more detailed local information such as that provided by the PILOT study (covered in the section relating to logistics, infrastructure and geomatics) to determine whether each option is technically and physically possible.

When the options have been refined sufficiently to identify the process fluid gathering, processing and/or product transfer and export needs, the Operations Representative (OR&A Manager/Advisor) can assist to determine how this can be facilitated and identify the necessary Operations Requirements for the operation and maintenance of each option.

At this stage, a preliminary estimate of CAPEX and OPEX for each option will provide key information to assist in the commercial evaluation of whether the proposed concept is viable.

It is likely that the estimate for OPEX will include a basic Flow Assurance study and the likely cost of any chemical additives required.

When a preferred concept is selected, the Operations Representative (OR&A Manager/Advisor) must be involved in the design process, especially for those high level decisions about materials selection, operating modes, availability, maintainability and operability.

Later in the project, as the project moves into the design stage (FEED) appropriate codes and standards must be agreed which will determine the selection of appropriate materials.

The Front End Engineering Design (FEED) must identify the Operations Requirements for operation, maintenance, pigging, inspection and corrosion management of the pipelines and manifolds.

The FEED should also address the safety related issues such as isolation, venting, purging, inerting and draining of the hydrocarbon inventory both in a normal planned context and in the event of an emergency.

A decision must also be made on the requirements for corrosion monitoring, pipeline integrity monitoring and leak detection systems. This will in all likelihood be directly proportionate to the magnitude of the asset and the risks related to oil spill issues.

Furthermore, because the pipelines and manifolds are often installed by a sub-contractor, the OR&A Team must ensure that the vendor and supplier data pertaining to the pipelines and manifolds components and equipment are included in the Asset Integrity Plan and Asset Register and that spares, consumables and other relevant data is captured in the Computerised Maintenance Management System (CMMS).

Where new technologies or systems are employed, training of the operations team must be identified and scheduled prior to the introduction of hydrocarbons.

Typical Deliverables/Outputs
- Opportunities and constraints which apply to the Operations scope and objectives.
- Operations contribution to routing/layout plans for pipelines and/or gathering systems.
- Pipeline Integrity Monitoring Philosophy, Strategy & Plan.
- Pipeline Leak Detection Philosophy, Strategy & Plan.
- Pipeline Corrosion Monitoring Philosophy, Strategy & Plan.
- Pipeline & Manifold Operating & Maintenance Philosophy & Strategy.
- Inclusion in project Asset Register.
- Spares, consumables and other relevant data captured in the Computerised Maintenance Management System (CMMS).
- Commissioning, Operating, Maintenance and intervention procedures.

6.2.6 Sub-Sea Systems (2.06)

Where the opportunity involves whole or partial location in an offshore environment, most of the issues should be the same as for the previous subject (Infield & Export Pipelines & Manifolds) though the fact that some or all of these facilities may be underwater adds a further level of complexity in addition to the reduced accessibility.

Participation in a PILOT (Preliminary Infrastructure, Logistics, Opportunities & Threats) Study of the Sub Sea Development aspects will determine the extent and magnitude of additional issues and concerns due to the marine environment.

Flow assurance issues are likely to be even more of an issue with sub-sea facilities, as will the mitigation of those issues.

The purpose of this subject in respect of OR&A is then to ensure that the selection criteria (including the operational requirements) for the sub-sea installations, the concept selection process and the design of the sub-sea facilities take into account the Operability, Maintainability and Availability issues identified in the Operations Requirements.

Typically, an early draft of the design of the facilities will be produced by a dedicated Sub-Sea Engineering team when an integrated view is difficult to form due to the limited definition of the entire facilities during the early stages of the project.

This makes it even more important for the OR&A engineer to identify any key Operability, Maintainability and Availability issues that may impact the well operability and ensure that these are specified in the Operations Philosophy and carried forward from the SELECT phase into the project and Sub Sea Installation Basis for Design (BfD) documents.

Additionally, operability of sub-sea installations that also utilise new technologies regardless of the topside being manned or unmanned (e.g. with increased levels of data acquisition and control capability) must be reviewed in order to create awareness of the increased level of capability the production system concept may be required to handle, within the project and operations teams.

A process to fully appraise any new technologies should be started in the SELECT phase since operability of the asset (resulting in increased/decreased cost/schedule of the development) may be significantly impacted by decisions made during this phase. Similar definition of each concept will be required regardless of whether the development is a Greenfield or Brownfield installation.

Typical Deliverables/Outputs

- Defined sub-sea installation content included in the detailed Operations Requirements.
- Flow Assurance plan(s) and studies related to sub-sea installation reviewed (e.g. Flow capacities of the Sub Sea Installation)
- Risks/opportunities around new sub-sea installation technology are fully captured in the Risk & Opportunity Management Plan.
- Endorsement of the early concept sub-sea installation design is made by the OR&A team.

6.2.7 Maintenance & Integrity Management (2.07)

A significant part of the Operations Readiness & Assurance (OR&A) Process is dedicated to establishing the Maintenance & Integrity (M&I) requirements of the asset during the project process in preparation for long term operation.

M&I Philosophy & Strategy

During the concept selection phase, the Operations Requirements are intended to identify and assess the key issues from a Maintenance and Integrity (M&I) viewpoint, that will affect future M&I activities in the OPERATE phase.

The M&I Philosophy and the M&I Strategy are a key components of this work and express the intended modes by which the asset will be maintained, services and repaired and also how the Asset Integrity of the facilities will be preserved during the OPERATE phase.

The M&I philosophy is a key part of the Operations Philosophy and at this stage, has a significant influence on its content, especially in respect of the intended availability of the facilities, which will in turn have a major influence on the following:

- Manning Levels, Shift/Rotation cycles & Accommodation,
- Physical Locations & Geomatics,
- Transport facilities (road, rail, air, sea),
- Local Infrastructure,
- Local Facilities,
- Support Systems,
- Safety Systems,
- Facilities for Health and Welfare.

Responsibility for many of these is shared with other disciplines such as the Logistics & Infrastructure team.

A structured maintenance system is critical to the maintenance of the asset and to maintaining Asset Integrity and all activities must comply with national and international laws and standards, corporate and local directives, rules, policies and processes for the maintenance and operation of the production and processing facilities and supporting infrastructure.

A number of options for 'maintenance strategies' exist and though these comprise only two main activities, namely planned and unplanned maintenance, the cost of the two can differ widely when the cost of implementing that maintenance (or not) is considered.

Planned Maintenance
The main types of maintenance are:

- Scheduled maintenance,
- Condition based maintenance and
- Breakdown maintenance.

For condition based maintenance, (planned maintenance based on equipment condition), condition monitoring facilities must be incorporated into the design of the equipment, together with a method for recording, reporting, analysing and planning maintenance activities.

These techniques also provide various options to refine the planned maintenance activities and include:

- **RCM** (Reliability Controlled Maintenance, the technique typically used on oil & gas processing facilities),
- **RBI** (Reliability Based Maintenance, normally used on systems where a loss of containment may be involved, (i.e. static equipment, vessels, piping or Pressure Safety Valves (PSV),
- **SIL** (Safety Integrity Level techniques are used for safeguarding systems such as Relief Valves and Fire & Gas detection systems).

These basic techniques offer a means of determining the consequence and probability of any failure.

No matter what particular technique is used, determining Risk (or the likelihood of a Risk occurring, basically a combination of consequence and probability) is the same across all techniques.

When conducting Risk Assessments, a Risk Assessment Matrix (RAM) is invariably used. This often includes some form of Failure Mode & Effects Analysis (FMEA) for critical components or items of equipment.

The generic term for these studies is Reliability, Operability, Maintainability (ROM) Engineering, intended where possible, to quantify the consequences of quality failures and to allocate Quality Assurance employing analytical techniques such as:

- HAZOP (Hazard and Operability studies)
- FMEA (Failure Modes and Effects Analysis)
- FTA (Fault Tree Analysis)
- ETA (Event Tree Analysis)
- RA (Risk Analysis)

Unplanned Maintenance
Unplanned maintenance is the term used to describe repair after an unexpected breakdown or failure.

This is often where significant additional cost components for OPEX is added if the equipment concerned fails to operate to the design intent. The effects of this can be disastrous if this affects the ability of the asset do satisfy the various contractual obligations of the asset (i.e. failure to deliver a pre-determined Daily Contractual Quota (DCQ)) which may also include penalty payments to customers.

A Breakdown Maintenance Strategy is usually only appropriate in the case of minor equipment where multiple units are installed or when the costs are known in advance (including down-time and lost availability). These can then be subjected to a Risk study and the outcome expressed in the form of a Risk Assessment Matrix (RAM) to determine whether the cost is considered to represent an acceptable level of risk.

M&I Activities

OR&A has a responsibility to ensure that the Operations & Maintenance team is capable of safe and sustained operation and maintenance of the asset during the OPERATE phase.

The OR&A team must therefore be involved from the early stages when the M&I Philosophy and Strategy is being developed, but becomes even more involved when ensuring the information to allow them to carry out the necessary activities is provided by the project team, vendors and suppliers.

The OR&A team must contribute to the Basis for Design (BfD) and the Invitation to Tender (ITT) and ensure the M&I related Operations Requirements are included.

In the later stages of the project, the M&I activities extend into reviewing the design information, participating in HAZOP, HAZID and HFE studies and in FAT and SAT activities. They are also involved in ensuring the necessary systems and equipment will be available for performing the various M&I activities including, but not limited to:

- Spare parts philosophy & strategy,
- Cost estimates for M&I activities,
- Testing and calibration facilities,
- Vendor/Supplier data,
- M&I software & systems development,
- Maintenance organisation development.

This usually culminates in the M&I team being involved in Commissioning and Start-Up (CSU) and performance testing activities together with the EPC contractor personnel before taking part in the final handover of the completed and tested asset to the Asset Operations Team.

During the early part of the EXECUTE phase of the project, when the maintenance organisation is established, the OR&A team will also be involved in identifying, sourcing and participating in the necessary training for Asset Operations Team personnel. This includes the Competence Assessment and Assurance activities.

The outcome of this part of the OR&A process includes confirming that:

- The Technical Integrity of the facilities is established and can be verifiably demonstrated at any time during the asset life cycle, (by monitoring the Design Integrity and Construction Integrity through the project phases),

- A Management of Change (MoC) process is implemented by the project to ensure decisions made during the Design & Procurement process are auditable,

- The availability & reliability of the facilities meet the Operations Requirements and design parameters and that this is re-confirmed at each stage of the design and construction process,

- Appropriate detailed design information, as specified in the ITT and BfD, is provided to the Operations team in a timely manner, to allow review and comment,

- The Computerised Maintenance Management System (CMMS) must be operational on receipt of the first items of equipment to allow preservation, integrity testing, commissioning (incl. SAT) & maintenance requirements can be planned and that baseline data can be established,

- The CMMS must be fully populated by the time the Project reaches the Ready for Start-Up (RFSU) stage,

- The Operations team is available, capable and competent to operate and maintain the facility to meet the promises made at FID, including the completion of necessary training, documentation and support services.

Asset Integrity

The involvement of the OR&A team is essential to provide assurance to the stakeholder that a clear and logical method of establishing and maintaining Design Integrity and the eventual achievement of overall Asset Integrity is in use by the project team.

Design Integrity sets the standards which are implemented through each phase of the project (involving the demonstration of Technical Integrity during EXECUTE) which will eventually lead to Operating Integrity (in the OPERATE phase).

This is the baseline which allows the Asset Integrity Management System (AIMS) to be used to manage and maintain Asset Integrity over the design life of the asset.

Most recently, some operators have included Process Safety Management as a component part of Asset Integrity systems, such that the various critical components (Safety Critical Elements) are managed by the same robust system.

Safety Critical Elements or SCE as they are commonly known must be subjected to a Written Scheme of Inspection and this is now a legal requirement in some countries.

The outcome of the early work in any project in respect of Asset Integrity begins with the creation of an Asset Register and the Asset Integrity Plan which contains details on each of the operable or maintainable components in the design and forms the basis of the information which is also input into the Computerised Maintenance Management System (CMMS).

There are many very good examples of proprietary CMMS systems available, such as Maximo, Datastream and SAP PM to name a few.

However, the key to a good system is not the name of the system, it is the validity and completeness of the information it contains and therefore the OR&A team is responsible for ensuring the CMMS system is fully populated prior to the introduction of hydrocarbons and that the system is up and running in time for the historical data from the start-up, commissioning and performance testing can be recorded as a baseline.

6.2.8 Process Automation & Control Systems (2.08)

Design Philosophy, Strategy & Plan

The purpose of this discipline is to ensure that a 'fit for purpose' process automation and control system is designed which closely matches the needs of the operations team in running the asset in a safe, sustainable and environmentally friendly manner for the duration of the design life of the asset.

During the concept selection phase, the initial Control & Automation Philosophy, Strategy and Plan must identify the key needs for Automation and Control for each of the proposed concepts, ensuring at all times that they fit with the agreed Operations Philosophy and Maintenance Philosophy for the asset.

The Control & Automation Philosophy will contain several chapters relating to various aspects of Control & Automation, including;

- Instrument Safeguarding,
- Fire & Gas Detection,
- Protection and Alarms Management,
- Life Cycle Costs,
- Selection and Positioning of equipment,
- Prevailing environmental conditions and equipment suitability,
- Human Factors Engineering.

In the early stages, these will simply be statements of intent which must be explored in detail only for the preferred concept.

The Control & Instrumentation requirements should also include reference to Flawless Project Delivery techniques and the intention to achieve and maintain Operations Excellence.

Design Considerations

During the detailed design phase, frequent reference must be made to the Control & Automation Philosophy, Strategy and Plan to ensure that the final product reflects the initial design intent. The main considerations are:

- Operability,
- Maintainability,
- Sustainability.

The temptation to include new technologies and/or over complication should be avoided (unless the benefits are significant) especially in harsh climates and areas where technical intervention may be difficult.

Operability
The final Control & Automation system must satisfy four criteria that are vital for the OPERATE phase:

- The asset can be operated safely,
- The asset operates to the design intent,
- Operation of the asset is sustainable,
- OPEX costs match the estimates, within acceptable limits.

Operability must be considered from both central and local controls, both of which should be designed to minimise the possibility of mis-operation leading to harmful consequences, (particularly in the case of the Safety Shutdown System).

Recovery from or mitigation of every conceivable abnormal operating condition (so far as is reasonably practicable), including transient conditions should be considered in the design to ensure the system is truly 'fit for purpose'.

Prior to start-up, detailed operating procedures must be prepared, operators (particularly control room personnel) must be fully trained and their competency assured.

Where feasible, a training simulator should be installed to allow frequent refresher training of operations and maintenance personnel. Where cost is an issue, the simulator can often serve as an off-line engineering interface for software maintenance or modifications, testing suspect components and also a source of 'hot spares' to further justify its inclusion.

Maintainability

The Control & Automation system should be designed taking maintainability into consideration as this could seriously affect the Life-cycle operating costs.

The system design must include built in test and diagnostic facilities with fault indication at the modular level. It should also allow a faulty module to be replaced while the system is live without causing further process disturbances.

In locations with restricted access, such as on offshore facilities, studies should be carried out to confirm the routes and methods by which equipment can be removed for off-site repair and subsequently replaced, including the provision of additional or temporary lifting equipment to aid such maintenance.

Sustainability

To ensure the sustained availability of the Automation and Control Systems, appropriate targets need to be derived for availability and reliability that must be included in the specifications for the systems. This may require an availability study for larger projects.

Where there is an issue with a major equipment item, provision of spare unit or inclusion of extra capacity in the design may be justified when comparing the total installed cost of a spare part against the cost of the effect on overall plant availability due to failure.

System Purpose
The stated purpose of the Process Control System is to allow the safe and efficient control of the plant, provide automatic protection to personnel, plant and equipment where appropriate, to assist in prevention of the environment and to minimise process downtime by facilitating prompt remedial action to correct abnormal conditions in the process.

The system must be capable of achieving the design intent by:

- Automatically sensing process conditions and operation of equipment, making minor incremental adjustments to maintain the process with the normal operating envelope,

- Automatically responding to signals from the Safety Shutdown and the Fire & Gas systems, bringing the process to a safe condition on receipt of shutdown signal,

- Providing the necessary control functions (and manual overrides) for the operation of plant and equipment from both central and local interfaces,

- Providing visual and audible information to enable the Operator to assess the equipment status and to attract the attention of the Operator to any event(s) which may require intervention,

- Providing facilities for the recording of alarms and key events,

- Providing facilities for trend displays of selected parameters,

- Providing HMI interfaces to the Safety Shutdown system, the Fire & Gas system and other third party control and monitoring systems.

Safety Critical Elements

A Safety Critical Element (SCE) is defined as a component or system which is used to prevent an abnormal condition or facilitate mitigating or remedial action. An SCE may be a single component or a system comprising a number of components.

Since the Process Control System and the Safety Shutdown System are integral to the implementation of such actions, they are also likely to be considered as Safety Critical Elements (at least in part).

The inspection and testing of SCEs is subject to specific legal requirements in many countries where a Written Scheme of Examination is required and stringent records must be kept.

Technology Risks

The temptation to include new technologies and/or over complication is strong in many design teams and this should be avoided (unless there are significant clearly proven benefits).

Over-complicated equipment and new technologies often suffer from increased maintenance requirements in harsher climates. Access times to the location for technical intervention may also be excessive involving increased down-time, (this is more evident when considering the availability and skill levels of local intervention teams).

If the decision is taken to include new technologies and complicated equipment, it is particularly important that the consequences of failure or delay in the procurement of the new technology should be fully accounted for in the project execution plan and project economics.

Design Documentation Requirements

The following project documents should typically be produced to cover, as a minimum, the design phases of the Process Automation and Control System:

Front End Engineering Design (FEED):

- Detailed Instrumentation & Control philosophy,
- Detailed Instrumentation & Control Specification,
- Block diagrams,
- I/O schedule,
- Process & Instrument Diagrams (P&ID),
- System architecture drawings

Detail Design:

As for FEED, plus the following:

- Detailed Instrumentation & Control philosophy,
- Detailed Instrumentation & Control Specification,
- I/O Schedule,
- Alarm and Trip Schedules,
- General Arrangement Drawings,
- Layout Drawings,
- Block Diagrams,
- Termination Diagrams,
- Loop Drawings,
- Process & Instrument Diagrams (P&ID),
- Data Sheets,
- Hook-up Diagrams,
- Hardware functional design specification,
- Software functional design specification,
- System schematic drawings.

All complex controls that comprise more than a simple cascade loop shall have a written control narrative describing the control functions.

All logic controls shall have cause and effect diagrams and/or function block logic diagrams describing and illustrating the logic functions together with a written description of the functionality. Sequenced functions shall also have a sequence chart.

6.2.9 Infrastructure, Logistics & Geomatics (2.09)
First Steps

It is essential for any development to be aware of every aspect of a new site or location and for all issues to be recognised and addressed (or even perhaps, mitigated).

The first step should always be to perform a Desktop PILOT study (Preliminary Infrastructure, Logistics, Opportunities and Threats) to develop/review the logistics strategy based on the available information from various sources (without travelling to the location).

The desktop PILOT study is essential to identify how access to the location needs to be made in the first instance and which areas require further or in-depth consideration from the point of view of the Logistics team.

It is intended to identify Opportunities and Threats which could constitute an early showstopper or add significant delay/additional cost to the project due to an oversight or omission (e.g. earthquake zone, war zone, extreme seasonal weather events, prevalent diseases).

Information Services such as the Drum Cussac Country Risk Reports available on the internet provide the latest security and travel risk related information and analysis on 200 countries and territories worldwide for travellers, analysts, risk managers and investors alike.

A great deal of in-depth information is available on the internet from many sources including official national and government sponsored information websites, tourism sites and unofficial or private websites. Both Google Earth and NASA websites are also available to the general public providing a wealth of Geomatics information in the form of accurate annotated maps and satellite imagery.

The combination of information available means that it is relatively simple to create a comprehensive picture of the proposed project location from an internet connected computer.

When a preferred location is identified, the desktop study can be followed up with a detailed on-site PILOT study to confirm and validate the information collected during the desktop study.

All of this information is absolutely essential information to the Concept Selection Team.

PILOT Study Considerations

A number of important considerations need to be addressed in the first instance, to explore the viability of conducting a project in the preferred location. Key Aspects of any PILOT study should include (but not be limited to):

- Geomatics, including:
    - Geographic location and access,
    - Local geography and site orientation,
    - Site elevation,
    - Climate (including winds, tides, precipitation etc.).
- Infrastructure
    - Native population (and the possible effects of the project),
    - Local Law enforcement,
    - Local import export laws, regulations, issues,
    - Local emergency facilities (Fire, Ambulance, Medical),
    - Local facilities & resources,
      (hotels, communications, finance, entertainment),
    - Local service companies, skills & manpower,
    - Neighbouring assets.
- Logistics
    - Local transport facilities and infrastructure,
      (Road, Rail, Air and Sea transportation facilities),
    - Site access (for construction traffic),
    - Basic cost estimates for Logistics activities.

Obviously, if there is more than one development concept under consideration, there are likely to be a number of differences in the logistical and infrastructure requirements for each.

To keep control of the preliminary work done, it is recommended that the desktop study be limited to a small number of possible concepts, listing the common requirements and the specific requirements separately.

The outcome of the desktop PILOT study should provide a detailed report with a list of selection criteria to enable the concept selection team to make an informed decision.

The first decision of the concept selection team should be to identify a preferred concept whereupon an on-site desktop study can be performed to confirm the findings of the desktop study and provide more recent data for the concept selection team.

Later Phases

As the project moves into the design phase, a Logistics Strategy can be developed for the selected concept. This must be cross checked against the corporate policies and philosophies of the Asset Owner to ensure it complies with the Company policies, rules, regulations and business processes.

Further checks must be made to ensure that proposed Logistics activities comply with local, national and international laws, rules and regulations.

A Logistics and Infrastructure Plan can then be developed which addresses the DEFINE, EXECUTE and OPERATE phases (since each of these will have very different requirements).

In the EXECUTE phase, the mobilisation of personnel and the provision of the necessary support services in Project locations will be paramount. This is more important if this requires the creation of a Logistics infrastructure where none is in place and adequate preparation time must be factored into the design. Other considerations should include:

- Acquisition of L&I services and equipment,
- Establishing forward supply bases,
- Supply of equipment and materials to site,
- Materials and Equipment Handling facilities,
- Secure transit, lay-down and transfer areas for materials,
- Suitable storage locations for materials and spare parts,
- Inventory Control and Materials Management systems,
- Logistics Management System,
- Transition plan to OPERATE phase.

During the OPERATE phase, after the demobilisation of the project team, the Logistics and Infrastructure will need to be organised, adequately manned and resourced to ensure the smooth running of the asset.

This will require the necessary Logistics systems to be in place and fully functional prior to the introduction of hydrocarbons.

6.2.10 Materials Procurement & Management (2.10)

Introduction

The purpose of this discipline is to ensure that a 'fit for purpose' process for Procurement & Materials Management is implemented which closely matches the needs of the operations team in running the asset in a safe, sustainable and environmentally friendly manner for the duration of the design life of the asset.

During the early project phases, most of the Procurement & Materials Management activities will be undertaken by the EPC Contractor so the involvement of OR&A is limited to ensuring that their Procurement & Materials Management system is indeed 'fit-for purpose'.

It is also the responsibility of the OR&A Team to ensure that the equipment procured satisfies the Operations Requirements expressed in the various design documents and that advance information on the Operability, Maintainability and Availability of the equipment is made available to the Operations team as soon as possible.

In the latter stages of the project, the OR&A team needs to identify the parts of the EPC Procurement & Materials Management systems that must be incorporated, maintained or adopted by the Asset Team for the OPERATE phase to ensure the continued availability of consumables, spare parts and other services from the suppliers/vendors.

Early Project Phases

During the concept selection phase, an initial Procurement & Materials Management Philosophy, Strategy and Plan will identify the key needs for Procurement & Materials Management for each of the proposed concepts, ensuring at all times that they fit with the agreed Operations Philosophy and Maintenance Philosophy for the asset.

The OR&A Team should contribute to identification of procurement threats and opportunities and their impact on each concept being considered, specifically identifying any advantages or disadvantages that would exist for each concept.

The procurement threats and opportunities and how they would impact on each of the concepts being considered need to be clearly understood as this will form part of the operations assessment (if one is being prepared for each concept that is being considered).

The OR&A Team will also contribute to development of a Procurement & Materials Management Philosophy and Strategy (usually as part of the operations philosophy for the selected concept).

The Procurement & Materials Management Philosophy and Strategy is required to:

- Set the foundation from which more detailed operations procurement plans are developed in later phases of the project.
- Support other OR&A activities (preparing estimates for manning levels, OPEX estimates, workshop requirements etc.).
- Identify where Procurement & Materials Management requirements can be included in CAPEX equipment purchase orders (e.g. initial spares purchase, spares holding agreements, after sales service etc.).

Design Phase
During the DEFINE phase, the OR&A Team contributes to the identification of Operations Requirements that must to be delivered by the FEED contractor and included in the Scope of Work section of the Invitation to Tender (ITT) document for the project.

It is essential that all Operations Requirements that are required to be delivered by the FEED contractor must be included in the Scope of Work, else they will not be included in the contractors bid and therefore would be unlikely to be delivered without additional costs.

Construction Phase
During the EXECUTE phase, especially during Detailed Design activities, the OR&A Team will be involved to ensure that the necessary information that will be required during the OPERATE phase is collected by the Project Team.

The information required for the OPERATE phase includes the necessary lists of spares should include commissioning spares, insurance spares and recommendations for operating spares for the first one or two years operations and estimated initial max/min stock levels.

It is essential for efficient maintenance and repair activities for detailed spare parts data to be available from the materials management system so that the initial spares delivery can be controlled and on-going re-ordering can take place.

The Procurement & Materials Management system used by the Project Team must be able to collate the spare parts data supplied by the various vendors and suppliers in a format suitable for uploading into the Computerised Maintenance Management System (CMMS).

The procurement team must put in place a transition plan for project delivery procurement agreements or contracts that need to be carried over into the OPERATE phase, including warranties and other agreed obligations. This activity should be conducted in accordance with the Project Handover Procedures.

The Transition Plan must identify and secure the procurement specialist support required to execute the Procurement & Materials Management related activities in the OR&A plan for the OPERATE phase of the project and include this in the project budget.

The OR&A Team will also identify the necessary training required to ensure the Operations Team are competent to operate the Materials Management System prior to the introduction of hydrocarbons into the asset.

6.2.11 IM, IT & Communications Systems (2.11)

Introduction

The purpose of this discipline is to ensure that 'fit for purpose' processes for IM, IT and Communications are implemented which closely match the needs of the operations team in running the asset in a safe, sustainable and environmentally friendly manner for the duration of the design life of the asset.

During the early project phases, most of the IM, IT and Communications activities will be undertaken by the EPC Contractor so the involvement of OR&A is limited to ensuring that the systems designs are 'fit-for purpose'.

Care must be exercised when planning the information systems needed to support a project and early consideration needs to be given to the overall IT Strategy moving into the OPERATE phase.

During the project phases before the asset facilities are populated, most of the IM, IT and Communications requirements will be provided by the EPC contractor or from the Asset Owner's organisation. This discipline is particularly concerned that the necessary facilities and equipment for the long term operation of the asset are included in the design.

In planning the IM, IT and Telecoms systems needed to support the operations team in the OPERATE phase, it is important that early consideration be given to the needs of the Operations Team and the asset.

Dependent upon the location and specific circumstances of the asset, the Operations team will have specific needs to enable them to operate the asset in a safe and sustainable manner, such as standard telecommunications facilities, Internet and data connections.

Considerations should also identify the anticipated number of users on the completed asset and the various hardware and software that may be required to satisfy these requirements, typically:

- Satellite communications equipment,
- IT Data Network systems (incl. servers & network),
- TV, projectors and other display equipment
- Telecommunications Hardware
- Specialist communications equipment such as satellite telephones, marine band radio and civil aviation band radio may also be required for remote locations.
- Standard business software and non-standard applications (e.g. eDMS, CMMS, @RISK, condition monitoring software etc.).

Early Project Phases

During the concept selection phase, an initial assessment of the IM, IT and Communications requirements for the OPERATE phase (and specifically at what stage they must be available for use) must be made for each of the proposed concept.

This will lead to the development of an IM, IT and Communications Strategy ensuring at all times that it complies with the agreed Operations Philosophy and Maintenance Philosophy for the asset.

In determining the IM, IT and Communications requirements for the asset, considerations should include:

- Appraisal and selection of appropriate computer systems to handle the large quantity of data and documents produced by the project;
- Preference to adopting systems which already integrate with the current systems in use corporately by the asset owner such as Accounting, Procurement and Logistics systems and Electronic Document Management Systems (EDMS);
- Encouraging the use of technologies for inter-site communication (e.g. voice, video, e-mail, and data);

Design Phase

During the DEFINE phase, the OR&A Team contributes to the identification of Operations Requirements pertaining to IM, IT and Communications that must be included in the Scope of Work section of the Invitation to Tender (ITT) document for the project.

It is essential that all Operations Requirements that are included in the Scope of Work of the ITT, else they will not be included in the contractors bid and therefore would be unlikely to be delivered without incurring additional costs.

The IM, IT and Communications will be set out in the Project IM, IT and Communications Plan, which should be an integral part of the Project Execution Plan.

It is usual for the Project IM-IT Team to develop a block diagram/flowchart to illustrate the proposed IM, IT and Communications systems at a basic functional level.

Construction Phase

The IT Strategy & Plan will be kept up to date with the latest Project developments and should therefore be regularly updated. A great deal more information will be available at this point in the project so more detail can be included.

There will be specific types of data created and used within the project, some of which will be discarded at the end of the project, but some which must be retained for use by the Asset Team and it is this data which the OR&A Team is specifically interested in. The types of data that the Operations function may require include (but are not limited to):

- Design data (drawings, specifications);
- Procurement data: equipment material requisitions etc.;
- Equipment vendor data pertaining to maintenance/inspection;
- Data required by law, regulatory bodies or classification societies, (not directly relevant for operating and maintaining the facility);
- OPEX data.

Asset data is usually captured and stored in electronic databases such as contractor engineering and/or CAD databases and maintenance and production support databases.

The Project IM, IT and Telecom Strategy should also consider the various interfaces, software and hardware and the methodologies used to facilitate exchange of this information.

Permits, Licence Fees, Taxes and Constraints
In many countries it is common for certain types of broadcasting equipment to be licenced, taxed or to require a permit for operation or use of the equipment. Similarly, the equipment may have limits, restrictions or constraints placed on the use of that equipment. The OR&A Team should track these details in the Licences, Permits and Permissions (LPP) section of the OR&A system.

Service Level Agreements
Although in-house IT staff can probably service, maintain and repair most of the IT equipment in use, some items may require specialist maintenance, repair and servicing attention. The necessary services may require some form of service contract, so the OR&A Team should therefore track these details in the Contracted Services section of the OR&A system.

6.2.12 Contracted Services (2.12)

Introduction

The purpose of this discipline is to identify the services that will require some form of on-going contracted services are implemented which closely match the needs of the operations team in running the asset in a safe, sustainable and environmentally friendly manner for the duration of the design life of the asset.

Early Project Phases

During the concept selection phase, an initial assessment of requirements for contracted services in the OPERATE phase (and specifically at what stage they must be available for use) must be made for each of the proposed concepts.

At this stage it is essential to identify issues which may affect the long term operation of the completed asset and include this information in the contracting philosophy and strategy for the asset. In the first instance, the PILOT study should ascertain the presence of international oilfield contractors, service providers and equipment vendors in the geographical locations or region of the asset.

A full market survey will determine the presence, capability and potential for development of local contractors to support the project during the project delivery and operational phases. The results should be documented in the initial operations assessment.

This information would allow the risks and opportunities, advantages and/or disadvantages in respect of contracted services to be identified and taken into consideration in the assessment of the strategies and plans for the proposed concepts being considered.

The outsourcing philosophy will be expressed in the contracting strategies and plans (including the risks and how they would impact on each of the opportunities) needs to be fully understood. This would then allow an Operations Contracting Philosophy and high level strategy to be developed and refined during subsequent phases.

The initial Operations Contracting Philosophy must:

- Set the foundation for the operations contracting plan;
- Support OR&A activities;
- Identify where services can be included in CAPEX as part of equipment purchase orders (e.g. spares, after sales service, etc.).

Design Phase

During the DEFINE phase, the OR&A Team contributes to the identification of Operations Requirements pertaining to contracted services and where these will be required prior to the handover of the asset to the Operations Team, these must be included in the Scope of Work section of the Invitation to Tender (ITT) document for the project.

The Operations Contracting Plan should begin to identify any specialist support required to execute the operations with contract related activities in the OPERATE phase of the project and be include in the OPEX estimates for the Asset budget in the OPERATE phase.

If support from contracting specialists is required to deliver the OR&A activities in the EXECUTE phase, they must be identified, agreed and included in the project budget for OR&A activities.

Construction Phase

The Operations Contracting Plan must be regularly updated to ensure it is kept up to date with the latest Project developments. As more information becomes available, more detail can be included. It will be beneficial if synergies with the EPC or design and construction contracts enable a follow-on arrangement with those contracted service companies which need to be extended into the OPERATE phase can be negotiated.

A plan for the implementation of the Operations Contracting Plan should be ready early in the EXECUTE phase to secure the resources required prior to Start-Up, to put the contracts in place and ensure a smooth transition to the OPERATE phase (including the transfer of the necessary contract documents and data).

As the contracting strategy and plan mature, more accurate cost data will be available that will enable the OPEX estimate to be updated.

Contracted Services will typically include, but not be limited to:

- Accommodation, cooking, cleaning, laundry etc.
- Site cleaning and site maintenance,
- Specialist services such as Laboratory, Analysis etc.
- Equipment and Vehicle hire,

In some cases it is entirely possible that the Asset Owner will engage the services of a single contractor to operate the entire facility. However, this does not remove the responsibility of the owner to ensure that the asset is operated in a safe, sustainable and environmentally responsible manner.

6.3 Operations Organisation & Competency (3.0)

6.3.1 Operations Organisation (3.01)

Introduction

The purpose of this OR&A discipline is to identify the requirements and extent of the organisation required to operate the facilities in accordance with the Operations (and Maintenance) Philosophy for the full design life-cycle of the asset. The OR&A Team, as the Operations Representatives in the project, are responsible for the development of the Operations Organisation.

Early Project Phases

During the concept selection phase, an Initial Operations Assessment will be required to determine the possible modes of operation of each proposed concept.

At this stage, the Operations (and Maintenance) Strategy determined by the Asset Owner (Project Sponsor) and the preliminary Operations (and Maintenance) Philosophy will state the preferred method of operating and maintaining the asset in line with the asset owners extended organisation.

Until a concept is selected, the Operations Organisation for each of the proposed concepts will be limited to an Organisation Chart outlining basic experience and skill sets and estimated numbers of personnel required.

When the project selects a concept for development, that Operations Organisation can be refined and key roles and responsibilities for the OPERATE phase can begin to be defined.

The importance of the operating environment of the completed asset must not be underestimated when identifying key risks and constraints that must be considered when defining the organisational design objectives, structure and delivery. The PILOT studies carried out during this phase will provide valuable insight to what is possible and what may be challenging.

Early identification of those local internal/external influences, requirements, and constraints will enable early creation of an organisation model that will support the development of the project and a robust Operating Organisation.

The various design objectives related to the operating environment including whether it is an Operated or Non-Operated Venture, prevailing legislation, venture and business objectives will impact upon the decisions made about the Operations Organisation.

At this point, it is therefore essential that the following information is prepared to facilitate the making of informed decisions:

- Risks/constraints that could impact Organisational Design options,
- Venture value chain(s) and value driver(s) that could impact business performance/Organisational Design options,
- Organisational responsibilities for the operating asset and functional support structures,
- Organisational objectives and requirements.

Design Phase

During the DEFINE phase, the OR&A Team contributes to the development of an Operations Manpower Plan for the selected concept.

Based on the Operations Philosophy for the asset location, the initial Operations Manpower Plan, based on the organisational structure for the concept, will be developed so that implications of Asset and Functional delivery can be considered against resource requirements (manning levels, cost, etc.), for the selected concept.

The Operations Manpower Plan will detail the following:

- Location specific constraints such as staff rotation, shift patterns etc. (particularly for remote locations, offshore facilities etc.)
- Organisation structure, manning levels and implications, including compliance with Operations Philosophy,
- Matrix of reporting relationships, roles and responsibilities at each stage of the asset development,
- Cost implications associated with the organisational design.
- Verification of organisational design against design objectives.
- Implications of organisational design on system availability using a Risk Assessment Model (RAM).
- Preliminary benchmark of staff numbers for the organisational design.
- Incorporate cost implications to OPEX estimate (particularly for remote locations, offshore facilities etc. as this will have a big impact on OPEX).

Construction Phase

The Operations Manpower Plan must be regularly updated to ensure it is kept up to date with the latest Project developments. The content of the plan must address the following:

- Examples and experiences from other asset operators in same region/location,
- Lessons Learned and Best Practices from other assets,
- Comparison to manning recommendations from the EPC Contractor/Vendors/Suppliers,
- Involvement in FATs, SATs, Commissioning and Start-Up,
- Hands-on experience with asset equipment for Operations Team prior to introduction of hydrocarbons,
- Prevailing laws, regulations and rules applicable for the asset location.

Early in the EXECUTE phase, the Operations Manpower Plan must be agreed with the Asset Owner/Operator and involvement of the Asset Team in the Commissioning and Start-Up activities should also be discussed with the EPC Contractor, (further reference to these activities are covered under the CSU disciplines).

OPERATE phase

The final Operations Manpower Plan should address all of these issues and present a phased plan detailing the assignment, roles and responsibilities of all personnel during each successive phase of the development until full Operational responsibility for the asset is assumed by the Owner/Operator.

The operations organisation needs to sustain the Operations Organisation throughout the OPERATE phase, to ensure the capability of the Operations and Maintenance to operate the asset in a safe, sustainable and environmentally responsible manner is maintained.

6.3.2 Operations Training & Competency (3.02)

Introduction

The purpose of this OR&A discipline is to identify the requirements and extent of the training and assessment of the competency of each role holder within the organisation to ensure their capability to operate the facilities in the required safe, sustainable and environmentally responsible manner.

The OR&A Team, as the Operations Representatives in the project, are responsible for the identification, sourcing and development of the Operations Organisation and establishing them as a competent workforce.

Resourcing of this role should not be underestimated as it has a significant impact on the long term operation of the asset.

Design Phase

It is normal for an EPC Contractor, when designing the asset, to provide an indication of the necessary manpower and competencies required to operate their design to achieve the design intent and any specific asset or process may include certain unique competencies that need to be fulfilled for successful operation of the asset.

However, the OR&A Team must compare this to the underlying Operations, Maintenance & Manpower Philosophies and the other various policies and strategies of the Owner/Operator to ensure compliance.

Construction Phase

When the final Organisational Plan is agreed, early in the EXECUTE phase the various competencies and responsibilities will be assigned to each role. The OR&A Team will then be involved (with the Operations Manager for the asset and the Human Resources team) in identifying, recruitment and mobilisation of suitable candidates for each role.

The OR&A Team must then perform a gap analysis to identify any shortcomings in the skill sets and competencies of the incumbent manpower organisation and Develop a Training & Development Strategy, plans, programmes & procedures to fulfil any identified competency requirements and close potential gaps and ensure the necessary competencies are achieved prior to the beginning of the commissioning phase.

The training plans/programmes for the future operations organisation must enable it to ensure that it is fully capable to Start-Up, Operate and Shut Down the facility in a safe, sustainable and environmentally sensitive manner.

The Training Plan for the EXECUTE phase will need to integrate the training requirements and schedule for vendor equipment training as it is a critical factor in the training & competency development of all members of the future operations organisation.

Careful planning and co-ordination of the necessary vendor training and the project schedule is essential to maximise the 'one-off' opportunity, during the Pre-commissioning, Commissioning, Start-Up & Handover phases.

Embedding the Operations team members in the CSU Team allows Operations Team personnel to gain real hands-on experience from Vendor/Supplier experts whilst they are still on-site by participating in Pre-commissioning, Commissioning, Start-Up & Handover activities.

OPERATE phase
The operations organisation needs to sustain the development and training of its staff during the OPERATE phase, to allow for training of new staff to replace those lost by general staff turnover, promotion or retirement.

For this purpose any processes and procedures developed during the project phase need to be retained by operations before Handover and maintained for future use.

This training should also include on-going training for ancillary/secondary roles such as Fire Fighting or Safety Intervention roles (refer to HSSE).

6.4 HSS&E Management (4.0)
6.4.1 Health, Safety & Environmental (HSE) Mgmt. (4.01)

Introduction

The purpose of this OR&A discipline is to identify the requirements and extent of the involvement of the Asset Operations Team in the development of the HSSE capability for the new asset.

The OR&A Team, as the Operations Representatives embedded in the Project, must ensure that they are fully aligned with the Project HSSE Team and available to assist them to carry out all of the necessary activities to prepare the asset and the asset organisation prior to the introduction of hydrocarbons.

Early Project Phases

During the concept selection phase, the OR&A Lead must first identify the interfaces between HSSE and OR&A (Operations) which may impact on the Operations Requirements and the decisions made in selecting the appropriate opportunity to proceed to the next phase.

There are a significant number of areas where HSSE and Operations share the responsibilities for key HSSE activities and deliverables and these may differ depending on the concept selected for further development.

During the early project phases, options and alternatives are identified, explored and developed to produce a preferred concept. The feasibility of the preferred concept (both technically and commercially) is assessed and a business case is put forward. It is necessary to recognise where input, contributions or action from the OR&A Team is required to ensure the correct outcome is achieved from the concept selection process.

In respect of Operations Requirements and interaction with HSSE activities and deliverables required during the early phases, these include, but are not limited to:

- HSSE related input into the Project Execution Assessment (PEA) and Project Assurance Plan (PAP)
- Preliminary HSSE Assessment,
- Identification of HSSE and engineering philosophies, standards, codes and specification applicable to the project
- Stakeholder Engagement Strategy

Most of this activity occurs during the High Level Risk Analysis activities undertaken during the concept selection process.

The involvement of the OR&A Team in the early phases may include site visits (such as during the on-site PILOT study and the Environmental & Social Impact Assessment).

If this is the case, certain precautions should be taken to ensure the safety and security of the Operations personnel sent to the site location. At this early stage, when the viability of the opportunity is still being evaluated, it is essential to identify the possible responses to an emergency situation.

Such an emergency situation could involve (but is may not be limited to) security, community, medical or civil unrest and plans to handle the response should be considered.

An appraisal of the Local Health Facilities and Medical Emergency Response options should be carried out to ensure the health and welfare of early visitors to the site location.

It is likely that the desktop PILOT study report will provide sufficient data to evaluate the current state of the site location in respect of safety and security of visitors.

This information should be the first contribution to the preparation of a preliminary Emergency Response plan for the opportunity.

Other early involvement for the OR&A Team will include the preparation of an Environmental & Social Impact Assessment (ESIA) Screening Assessment of the proposed development. It should address activities or effects which may impact one or more of the following:

- Local Environment,
- Local Society,
- Local infrastructure,

These effects may include, but are not limited to:

- Additional traffic (cars, trucks, trains, aircraft),
- Noise, smells, litter etc.
- Leaks, spillages, discharges, hazardous substances, waste products, pollution,
- Expansion of local population,
- Effects on local resources (hotels, shops etc.)
- Effects of incidents involving company or contractor assets.

Design Phase
Once a preferred concept is selected and the Project moves into the DEFINE phase, the involvement of the OR&A Team in respect of the activities of the HSSE Team increases significantly.

It should be remembered at this point that the purpose of the involvement of the OR&A Team is to ensure that the necessary operations related issues are addressed during the design phase, to provide assurance to stakeholders that the necessary HSSE work related to the Operability, Maintainability and Integrity of the asset will be completed at the relevant stage in the project.

To this end, it is essential that the OR&A Team are involved in the following (as a minimum):

- Preliminary Hazard Identification Studies (HAZID),
- Development of Asset Risk Register,
- Development of preliminary Waste Management Plan,
- Contributions to the Environmental Management Plan,
- Participation in HSSE Safety Engineering Studies,
- Contribution to HFE Screening & HRA studies,
- Review of Vapour Cloud & Gas Dispersion Study,
- Contribution to HSSE requirements for Contractor(s),
- Contribution to Occupational Health Risk Assessment (HRA),
- Contribution to identification of Safety Critical Elements (SCE),
- Contribution to preliminary HAZOP,
- Contribution to HSE Case,
- Contribution to development of Permit to Work (PTW) System,
- Development of Simultaneous Operations Procedures (SIMOPS),
- Development of Combined Operations Procedures (CONOPS),
- Development of Manual of Permitted Operations (MoPO).

Construction Phase

As the Project moves into the EXECUTE phase, the involvement of the OR&A Team in respect of the activities of the HSSE Team increases further.

The involvement of the OR&A Team ensures that the operations related issues are included during the EXECUTE phase, providing assurance to stakeholders that the necessary HSSE work related to the Operability, Maintainability and Integrity of the asset will be completed as the relevant stage in the project passes.

The OR&A Team will be involved in the following (as a minimum):

- Hazard and Operability Study (HAZOP)
- Updating Asset Risk Register,
- Updating Waste Management Plan,
- Updating the Environmental Management Plan,
- Participation in HSSE Safety Engineering Studies,
- Contribution to Equipment/Site Layout,
- Updating HSSE requirements for Contractor(s),
- Contribution to Written Scheme of Inspection for Safety Critical Elements (SCE),
- Contribution to HSE Case,
- Implementation of Permit to Work (PTW) System,
- Implementation of Simultaneous Operations Procedures (SIMOPS),
- Implementation of Combined Operations Procedures (CONOPS),
- Implementation of Manual of Permitted Operations (MoPO),
- Full ALARP demonstration,
- Pre-Start-Up Audit (PSUA).

At all stages, the HSSE activities must comply with the stated HSSE Policy, Strategy and Plans of the Asset Owner/Operator.

Operate Phase
The HSSE Organisation needs to maintain Asset Integrity and Process Safety throughout the OPERATE phase and the systems, processes and procedures developed during the project phases are essential in preparing the OPERATE phase personnel for this task.

6.4.2 Security Management & Community Relations (4.0)

Introduction

The purpose of this OR&A discipline is to identify the Security requirements and extent of the necessary Community Relations of the Asset Operations Team with the various stakeholders both local and national.

Regulatory Regime

Although the PILOT study will provide a basis of the necessary information, it is essential to identify applicable security standards/guidelines affecting Operations activities of the proposed asset.

This should begin with the creation of a detailed list of all (Current) applicable regulations pertaining to the location of the asset. These regulations (company/local/regional/national rules) must then be examined to determine the extent to which they affect security and how they may affect the security related aspects of the proposed Operations activities.

The method of determining the extent to which Operational Safety may affect the Operations personnel includes participation in a Preliminary Security Assessment led by the HSSE (Security) team.

Prevailing Local Situation

The Preliminary Security Assessment should produce information about the safety and stability of the host country, the local security situation.

It should take into account the magnitude of the efforts that may be required to secure the safety of the Operations Team and any other OMV personnel in-country, including the possible intervention of or assistance from the local law enforcement infrastructure in any possible breach of security.

The assessment should take into identify possible threats to the safety of OMV personnel (cultural or otherwise) during normal operations and also the safety of those personnel during various emergency conditions.

The contribution from the Operations Team to facilitate an effective assessment to be conducted includes:

- Manpower Numbers;
- Rotations & Movements;
- Proposed Accommodation;
- Exposure/Risk Assessment in respect of:
    - Disease/Health Issues;
    - Crime,
    - Politics,
    - Religion,
- Gender based concerns;
- Community Relations;
- Tribal/Cultural Issues;
- Security of Communications.

Security Planning

The Preliminary Security Assessment will allow the concept with the most stable security environment in respect of Operations activities to be identified. This information will be taken into account during the concept selection process.

The Security Plan for the Project (which will eventually be developed into the Security Plan for the OPERATE phase) should led by the HSE Security team, the entire Project Management team should be involved in developing the Security Protocols and Procedures for the Project (and which will eventually be developed into the Security Protocols and Procedures for the OPERATE phase).

Business Risk Consultants such as Drum Cussac should be consulted as they can provide online Country Risk Reports with the latest security and travel risk related information and analysis on 200 countries and territories worldwide for travellers, analysts, risk managers and investors.

Business Continuity

Business Continuity is a key issue which must be considered, especially in high risk environments. The possibility must be addressed that an evacuation from the site (or even the Country) may be required at some point, in response to a difficult situation.

Procedures for Business Continuity must consider the practicalities of stopping whatever project or asset activities are being undertaken and moving the entire workforce to a safe location.

The Business Continuity Plan must consider a range of situations and scenarios where partial or total evacuation from the site, region or Country may be necessary (or even imposed by Government/Military action).

It should also consider the timescales involved in making the workplace safe (shutdown, mothballing etc.) and the consequences of that action in respect of the eventual return to site.

The Business Continuity Plan should also contain a section pertaining to remobilisation and recovery from such situations.

Early Project Phases

Obviously, during the concept selection phase the exposure to personnel is limited as there will only be a small number of personnel on-site at any point and this can often be managed (even in high risk areas) by engaging private security firms to provide comprehensive security cover.

However, as the project moves into the DEFINE phase, the number of personnel exposed to whatever risks are identified will increase significantly.

It is therefore essential that a comprehensive Security Policy, Security Plan and supporting procedures are prepared at the outset and are updated as the project progresses.

Construction Phase

As the Project moves into the EXECUTE phase, the requirement for security will increase exponentially (especially in high risk environments). It is essential that the Security team have already established a presence in the asset location and have built working relationships with the local people.

This early involvement must include an element of Community Relations and it is essential that the person (or persons) responsible for the point of contact with the local people is seen to have the respect and support of the Contractor Management Team and the Asset Owner/Operator.

Many of the inevitable issues which cause civil unrest in high risk environments can be defused by negotiation before they become unsafe. It is always a good idea to identify roles which can be staffed by local people which helps to build the necessary relationships.

Operate Phase

The HSSE Organisation needs to maintain Security and community relations throughout the OPERATE phase and the systems, processes and procedures developed during the project phases are essential in preparing the OPERATE phase personnel for this task.

6.4.3 Risk Management (4.03)

Introduction

The Management of Risk is a fundamental component of Project Management and managing risk efficiently can be the difference between success and failure of a Project. It should therefore be very high on the agenda of anyone who has a role in the decision-making hierarchy of a Project.

Risks are factors that could influence whether a project meets its business objectives. In Project terms, a risk is defined as the **likelihood** of an event occurring and the **impact** of that event on the ability of the Project (or Asset) to achieve those business objectives.

The initial proposal for any Oil & Gas development is based on a series of assumptions. Each of those assumptions, by their very nature, includes an element of Risk.

The assumptions concern high level issues such as business reputation, HSE and social performance, but as the project progresses, include project schedules, production rates, CAPEX and OPEX costs, price of the products, market values and the viability (operability and maintainability) of the new asset.

Risk outcomes are often considered to be negative events (threats), but the Risk Management process should also be applied to positive events (opportunities).

An area where the complexity and importance of risk to a company is increasing is their strategic approach to reputation management and the involvement of the External Affairs team (especially in the aftermath of an incident where the Crisis Management processes of the company will be scrutinised mercilessly).

The purpose of the Risk Management process is therefore to identify those risks early, ensure the relevant people are aware of them and manage them throughout the duration of the Project and the Life-Cycle of the Asset.

All Risk Management Systems (RMS) for a project requires effective communication to ensure that all of the risks at every level of the organisation are fully understood. To achieve this, specific software tools (e.g. EasyRisk) are required and the Risk Management Process must be a key component of the Project Management Process.

Risk Management Process

Managing risk effectively requires the implementation of a robust Risk Management Process that contains the following processes:

- **Identify** the Risks, identify the owners of the risks and record them in a Risk Register,
- **Assess** the severity of the risks in terms of likelihood and effects, including a method of prioritising those risks and focusing the responses to them,
- **Plan** appropriate responses to each of the risks identified using ALARP principles and record these in the Risk Register,
- **Implement** the planned responses as required, monitoring and appraising and recording the effectiveness of those responses,
- **Review** the identified risks frequently to reassess their severity and revise the response plan as required, including the close out of risks found to be no longer applicable,
- **Improve** the Risk Management process by ensuring the process is effective, (if the process is too onerous the system may become overloaded with minor risks whereas too lax a process may result in serious risks being overlooked).

Risk & Opportunity Management Plan

Prepare a Risk & Opportunity Management Plan for each concept. Develop and update this plan throughout the project to eventually arrive at a detailed plan for the OPERATE phase.

The purpose of this activity is to ensure a balanced decision as to which option is selected, weighing the value of each concept and the associated risks (upside as well as downside).

Detailed risk analyses may be required to evaluate the quantitative impact of the risks and enable quality decision-making. To ensure that the risks and opportunities of all the project's development options are identified and made manageable in a structured manner to aid selection of the optimal concept.

The identified risks should be displayed in some form of matrix which allows the relative priority to be assigned to each. This usually takes the form of a graph charting the relative likelihood of the occurrence of each risk against the consequences of the risk, should it occur.

Categorisation of Risks

A structure which is commonly used to categorise risks is known by the acronym TECOP. This acronym stands for Technical, Economic, Commercial, Organisational and Political and is a very good method of sorting risks at a high level into five generic categories.

Even though a risk may be logically placed in a particular category, it is important to recognise that a risk will rarely occur in only one category and it is also highly unlikely that it the impact will be seen in only one category. The table below provides some guidance on TECOP categories:

TECOP Category	Risks
Technical	Includes reservoir, wells, surface facilities, infrastructure, HSE, new technology, operability, maintainability and functional issues. Usually high on the agenda when Project risks are discussed.
Economic	Mainly financial risks in respect of CAPEX and OPEX or the effects of failure to meet scheduled milestones.
Commercial	Often divided into two aspects: **Contracting**: The use of contractor companies to provide goods and/or services and the associated risks, **Legal**: The laws, regulations and legal infrastructures of both the location and the customer for the product need to be considered as some are often in conflict. This may affect licence agreements and concessions.
Organisational	Usually associated with the organisational structures involved i.e. Project Team, Asset Owner, Contractor relationships, more likely to occur during the transition between project phases.
Political	Includes Social, Reputational, External Affairs, Communications, Media Relations and the effects of any or all of these on the local infrastructure and wider world.

Early Project Phases

During the early project phases, the OR&A involvement in Risk Management is to ensure that a robust Risk Management Process is implemented in the project and that a Risk Register for each concept under consideration is established.

At this stage, the main purpose of the Risk Management process will be to provide data for assessment and comparison purposes, used in the selection of the preferred concept for further development.

The main questions to be asked at this stage are:

- Have all the risks been identified?
  and,
- Has the best concept for future development been selected?

Design Phase

During the DEFINE phase, the Risk Management process focusses on identifying the risks associated with the EXECUTE and OPERATE phases at a more detailed level.

This may have a significant effect on the design process as risks are identified and the appropriate responses to each risk are identified and may involve the 'designing out' of key risks.

The most efficient response to any identified risk may be to Accept, Mitigate, Transfer or Remove that risk and this is far more easily achieved during the design stages.

The main question to be asked at this stage is:

- Is the Project ready to manage all the identified risks?

Construction Phase

By the time the Project reaches the EXECUTE phase, changes to the design to mitigate or remove risks becomes far more expensive and time consuming. However, the main risks at this stage revolve around achievement of scheduled milestones and delivery as promised.

This in itself can introduce problems if additional (unplanned) risks are taken in a misguided attempt to achieve a challenging milestone.

The main question to be asked at this stage is:

- Are all the identified risks being effectively managed?

Operate Phase
The OPERATE phase required continued management of the risks identified during the earlier phases, but will also introduce risks of its own. It is therefore essential that the Risk Management Process is managed carefully during the transition of ownership from the Project to the Operations Team and that none of the identified risks is overlooked.

Summary
There will be identifiable risks within every OR&A discipline, not least the risks inherent failing to implement the OR&A process from the earliest point in a Project (leading to a failure to provide the appropriate level of assurance to stakeholders).

All of these risks must be captured in the Project Risk Register and the appropriate risks carried forward into the OPERATE phase Risk Register.

A review of the Lessons Learned from earlier projects will demonstrate whether the OR&A Process was implemented at the appropriate juncture and this will support the timely implementation in future projects. However, this will only be the case if the Project Management Team understands the purpose of OR&A and the added value it brings to a Project.

6.5 Commissioning & Start-Up Management (5.0)

6.5.1 Commissioning & Start-Up (CSU) (5.01)

Introduction

The management of Pre-commissioning, Commissioning, Start-Up and Performance testing is a key component of the OR&A Process, purely because the Operations Team has significant involvement in every stage.

The CSU activities also interfaces with the Flawless Project Delivery component of OR&A since a smooth, first-time start-up and steady ramp-up to the design parameters is a key deliverable for both.

Early Project Phases

During the early project phases, the OR&A involvement in Risk Management is to ensure the task of managing the OR&A activities related to Pre-commissioning, Commissioning and Start-Up of the Project is usually assigned to an OR&A CSU Lead Engineer with extensive Commissioning and Start-Up experience.

At this early stage of the development this position will be part-time, probably consultation with a Subject Matter Expert (SME), particularly if the OR&A Lead is not suitably experienced in the principles of Commissioning and Start-Up (CSU).

The involvement of a CSU expert will be limited to high level decisions which will have a significant effect on the project activities, particularly the effect on Commissioning and Start-Up (CSU) activities.

A Preliminary Field Development Plan detailing issues and options for CSU delivery and an estimate of possible manpower requirements should be prepared for each concept under consideration.

This should also include a brief description of the necessary CSU activities for each concept, assessed by suitably competent, experienced and qualified CSU SME, including the necessary HSE related issues and should also consider Lessons learned from previous projects.

Failure to identify CSU risks and issues during the early phases, when selecting a preferred concept for further development, could detrimentally affect that concept and the ability to of the Project to commission and start up the facilities effectively, leading to serious financial repercussions to the project.

Design Phase

During the DEFINE phase, the CSU Lead will be a full-time position and he should be preparing the first draft of the Commissioning Plan. At this stage it is only a draft, and will identify the requirements of the Commissioning Team to allow them to efficiently commission and start-up the proposed facilities.

The commissioning requirements will form a part of the overall Operations Requirements to be included in the Basis for Design (BfD) and the Invitation to Tender (ITT) documents.

A preliminary Commissioning Strategy must be prepared for the selected concept to ensure all aspects of Pre-commissioning, Commissioning and Start-Up are assessed. It will define the requirements of the commissioning team will be based on the information available from the Front End Engineering Design (FEED) team at this stage.

The Commissioning Strategy will need to be refined as the design is improved and finalised and must be checked to ensure it complies with the overall Philosophies and Strategies of the Asset Owner and the Project.

Construction Phase

As the project moves into the EXECUTE phase, after the Financial Investment Decision (FID), the award of the EPC contract and the beginning of the Detailed Design work, the Commissioning Strategy will be updated and the Commissioning Plan will be expanded to include the following:

- Commissioning Management System (CMS),
- Commissioning Guidelines & Procedures,
- FAT/SAT Schedule (with HSSE),
- Commissioning Schedule (with EPC Contractor),
- Commissioning Manpower Plan
  (including involvement of the Operations Team),
- Interfaces with Flawless Project Delivery,
- Interfaces with Training & Competence Team (OR&A),
- Commissioning Spare Parts,
- Commissioning First Fills,
- Commissioning Consumables, Lubricants & Fuels,
- Commissioning Milestones,
- Early Production, SIMOPS, CONOPS,
- Performance Testing Criteria,
- Handover & Acceptance Plan,

The Integrated Project Plan (IPP) will list many of these items, but the detail must be agreed with the Commissioning Team. The Commissioning Plan must be prepared and agreed with the EPC Contractor and must include all of the requirements of OR&A, FPD and other Operations Requirements.

The Commissioning Management System (CMS) used by the EPC Contractor must be capable of providing detailed progress and completion status reports and of recording deviations and changes in the Commissioning Plan.

An electronic document management system (eDMS) compatible with that used by the client (Asset Owner) must be used by the EPC Contractor and Project Document Control Centre (DCC) to ensure an efficient method of control and handover of the necessary documentation is maintained.

The Commissioning Plan should define the milestones and the various work packs (systems and sub-systems) to be (pre)-commissioned in a clear and structured manner, accompanied by the necessary commissioning procedures.

Start-Up, Performance Testing & Operations
From immediately prior to the introduction of hydrocarbons to the asset, the responsibility for Operations belongs to the Operations Team (as the Licence Holder) and this can only commence when a Pre-Start-Up Audit (PSUA) has been successfully completed and any findings or recommendations from that audit have been fully implemented.

The PSUA will address the completion of pre-commissioning activities (include accepting facilities and equipment from construction by confirming mechanical completion and raising punch-lists where required), confirmation that the Asset Organisation is capable of starting (and stopping) the process in a safe, sustainable and environmentally sensitive manner. In short that the asset has reached a pre-determined level of '**Readiness to Operate**'.

The definitions of the appropriate milestones and stages of Start-Up, Performance Testing & Operations activities need to be clear and precise as it may not be the same as the optimal sequence for the construction contractor. These definitions and the method by which any deviation from the plan is controlled must also be clearly defined in the construction contract.

Detailed performance testing and acceptance criteria must also be specified in the contract with the EPC Contractor and included in the Handover and Acceptance Plan.

Decommissioning
It is a legal requirement in some countries that all Oil & Gas developments make reference to the eventual decommissioning of the facilities under construction.

However, many companies recognise this as a hidden cost for the future and go further than that requirement, requesting the EPC Contractor designing the facilities to include facilities for decommissioning in the design.

6.5.2 Asset Handover & Acceptance (5.02)

Introduction

The management of the Handover & Acceptance process for the asset is a key component of the OR&A Process, because this is the process by which the Responsibility for Operations and subsequently the Ownership of the Asset is transferred to the Operations Team whereupon they become accountable for the long term operation of the asset.

It is essential that all of the necessary brochures, data, drawings, procedures and instructions are retained is a system which allows the Operations Team to Operate and Maintain the asset in a safe, sustainable and environmentally sensitive manner for the full design life of the asset.

Early Project Phases

As soon as a clear preferred concept is identified for further development, it is necessary to begin the development of a Handover and Acceptance Strategy to ensure that handover requirements are driven by a consistent strategy throughout the phases of the project. This should comply with the stated philosophy of the Asset Owner.

Although this may seem to be very early days, it is necessary to identify any criteria or requirements related to Handover and Acceptance which may need to be included in the Basis for Design (BfD) and Invitation to Tender (ITT) documents. This ensures that the construction and commissioning teams working for the EPC(C) Contractor are fully aware of the handover requirements from the outset.

The Handover and Acceptance requirements should also be detailed in the Project Execution Plan (PEP), the Interface Engineer (Handover and Acceptance Lead) should be identified as the responsible person and focal point for all issues relating to handover.

Detailed performance testing and acceptance criteria (including the duration of the testing) must be specified in the contract with the EPC Contractor and included in the Handover and Acceptance Plan.

Design Phase
As the detail increases during the DEFINE phase, the requirements for Handover & Acceptance of the completed facilities will become far more detailed.

The Handover and Acceptance Plan must be frequently updated to ensure the latest details and revised requirements are incorporated in the BfD and ITT.

Construction Phase
From the start of the Detail Design activities, the CSU Lead must ensure that the necessary management controls, procedures and structure required to ensure successful handover, (including the planning and execution of the necessary activities) and acceptance of the asset by the Operations Team, must be implemented in accordance with a written Handover & Acceptance Manual and this should be specified in the PEP.

From a Quality Assurance perspective this should also include checks for completeness of as-built drawings, records and specifications, availability of vendor documentation for all equipment and inclusion of experience gained during commissioning into the operating and maintenance procedures (as-built). It should also specify the population of a 'fit for purpose CMMS' to facilitate the Operation and Maintenance of the asset.

Handover & Acceptance System
To ensure the handover process is methodical and complete, a clear process is required which identifies each of the necessary roles and responsibilities during the handover process.

The handover process begins when a Ready for Start-Up certificate is issued by the Project Owner (Asset Owner) after confirmation of Mechanical Completion, completion of the necessary pre-commissioning activities and the successful completion of a Pre-Start-Up Audit (PSUA).

Planning for the handover activities must be completed and implemented well in advance of this point. Detailed handover procedures must be prepared for each project function and activity to ensure all parties are fully aware of their duties, roles and responsibilities in respect of handover.

Each focal point must prepare the appropriate handover dossiers for each commissioned package completed, in accordance with the Handover Matrix.

Each handover dossier must then be formally accepted by the Asset Operations Team, using the agreed handover process, to ensure a full and complete handover is achieved.

Finally, a system for recording, quantifying and clearing outstanding punch list items must be implemented to identify any outstanding or incomplete work or items.

Prior to the introduction of the first hydrocarbons into the facility, the Responsibility to Operate must be formally transferred to and accepted by the Asset Operations Team.

Operate Phase

The final handover (of ownership of the facilities) cannot be completed until a successful performance testing programme is completed by the EPC(C) Contractor, so the running and operating of the facilities is performed by the CSU Team (which includes a large percentage of Operations Team Personnel).

The Operations Team personnel will continue to work alongside the commissioning team during start-up and performance testing.

During this period a continuous transfer of information from the EPC(C) Contractor to the Asset Operations Team will take place. The Handover Process is used to ensure this is achieved completely and accurately.

This must include a clear timetable to ensure all the appropriate data is received, in a timely manner).

Using this method, the Handover and Acceptance of the asset is fully documented and the status of each component, equipment, sub-system and system can be recorded.

Maintaining integrity of such a system becomes significantly easier when the data pertaining to every item is clearly documented is a system which allows the retrieval of the necessary records and data to facilitate the Operations and Maintenance activities, improving the move towards ensuring and maintaining Asset Integrity.

7 Implementing OR&A on a Project

7.1 Dissemination of Information

The first issue facing an OR&A practitioner on a Project is how to disseminate the vast amount of information required to implement OR&A in a way that neither 'swamps' users in a sea of information nor 'starves' them of vital information required to complete the various activities and tasks at the appropriate time.

The methodology to solve this problem was pioneered by OR&A Ltd. when creating an OR&A system for Shell, but since it uses only standard Microsoft Office applications and HTML pages, it can be used by anyone prepared to take the trouble to set it up.

The method deployed by OR&A Ltd. uses a number of web (HTML) pages to provide the functionality required to manage the volume of information within the system and to present it to the user in a manageable, context related fashion.

By creating an easily navigable yet robust structure, the user is presented with only a small number of choices on each web page, so with just a few clicks the user can select the appropriate Subject, Phase and Activity, gaining access to Task level and the detail needed to carry out the actual work.

Because of the use of hyperlink functionality, supporting information in PDF format is also made immediately accessible.

The important points to remember are that OR&A Ltd. has extensive experience in creating these systems, reducing the time and effort required, and that for any particular Company the system needs to be created only once.

By using this approach, a central point for the management of information is retained, minimising the maintenance requirements for the system yet ensuring that all users have access to the latest version of the information provided.

Note: The assurance software package supplied by OR&A Ltd. complements the information system so precisely that moving between the two is almost seamless.

7.2 Setting-Up the OR&A Application
7.2.1 Introduction
The desktop application used by OR&A Ltd. to implement the OR&A Process on a Project requires minimal user training and minimal set-up of the software, allowing the user to get on with using the system in a matter of minutes.

The OR&A desktop application, although easy to use, requires the user to understand a few key terms to enable its use to be clearly explained.

7.2.2 Definitions
Deviation
The OR&A System is a robust system that initially sets every Activity and every Task as a required deliverable. However, in most cases, there are certain Activities and Tasks that are not required for the current Project.

Because the OR&A application monitors the progress of every Activity and Task, (evaluating progress against the recorded data), the user must identify which activities or tasks are not required (and will subsequently not be monitored).

Identifying an Activity or Task as 'not required' is called 'Entering a Deviation', since it is specifically directing the application to deviate from the OR&A process by not monitoring that Activity or Task.

When a deviation has been entered against an Activity or Task, the progress of that Activity or Task is no longer factored into the calculations made by the application and therefore has no further effect on the indications of progress shown in the application screen views or on the printed output.

To ensure that only deviations that are authorised are entered against an Activity or Task, the application requires the approval of the appropriate authorities from a RACI Matrix to validate the deviation.

A hard copy of the entire list of deviations in the current Project, together with the details of the user entering the deviation, the accountable and responsible persons authorising the deviation and the time and date that each of these was entered can be printed at the touch of a button.

As with all robust database systems, this information cannot be deleted. The deviation can be removed at any time, but the fact that it was added then removed is automatically logged by the application.

This prevents the user from 'tweaking' the system to artificially produce a more favourable indication of project status.

RACI

The acronym 'RACI' is used to describe a chart that is used to identify the **R**esponsible or **A**ccountable persons or those who should be **C**onsulted or **I**nformed about a particular detail or action, hence the acronym **RACI**.

Within the OR&A System, such a RACI chart is used to illustrate a Matrix of Authorities for the OR&A Disciplines. These are usually the Project Manager, the OR&A Manager/Lead or the relevant Subject Matter Experts (SME) for a given group or discipline.

Not Ready to Begin (Too Early)

The OR&A Application displays Activity and Task status using a 'Traffic Light' style indication. The point of transition between colours can be adjusted by the user, but by default is set at 20% from Red to Amber and at 80% from Amber to Green.

To minimise congestion on the overview display screens, the value assigned to each Activity or Task is not displayed, only a simple coloured light is used to indicate the current status. The full information is available and can be easily viewed if required, on another more detailed screen or on the full OR&A Dashboard.

At the start of each project phase, the first Activity in each Subject (as shown on the Delivery Map for that subject on the OR&A web page) will be automatically set as 'Planned' and will be assigned zero progress/completion on the OR&A Dashboard. The traffic light will indicate RED.

Activities that are not yet in scope, (in other words that are not intended to be started in the immediate future), are set as 'Not Ready to Begin'. They will therefore be 'greyed out' on the OR&A Dashboard and will not be assigned a progress/completion value.

As each activity is set to 'Planned', the subsequent activity will be set as 'Not Started' and will be assigned zero completion on the OR&A Dashboard. The traffic light will indicate RED.

This functionality prevents the clutter that hundreds of tasks showing as 0% with a red Traffic Light would cause, when it is clearly too early to measure them.

8 Measuring and Monitoring OR&A

8.1 Recording OR&A Progress/Status

The entire content of the OR&A System, in respect of Groups, Subjects, Activities and Tasks can be easily represented on a spreadsheet. However, for even a relatively small project, the spreadsheet is likely to extend to some 2,500 line items.

The effort required in recording on-going status information in chronological order, amending completion status (by % and Traffic Light) and to collect other pertinent data for OR&A (such as recording status information or identifying potential lessons learned) on such a spreadsheet is prohibitive.

Similarly, the production of weekly or monthly reports using such a spreadsheet would be resource intensive and prone to errors in recording the data.

Production of a frequent 'Planning Report' which provides incremental updates to the Project Planner would also be a resource intensive activity with similar limitations to the weekly and monthly reports. Comparison of the actual progress recorded on the spreadsheet with the progress reported in the Integrated Project Plan would be a full-time job.

However, the system developed by OR&A Ltd. employs a client side software application which provides a simple solution to these problems.

The structure within the OR&A application mirrors that of the OR&A System and makes recording progress against Activities and Tasks even simpler than using a spreadsheet. Recording other additional status information is equally easy.

The dashboard style display used by the application allows the user to view and add information in 'real-time' and to see the effect this has on the overall implementation progress/status as the screen display is immediately updated.

8.1.1 Monitoring Progress

Effective implementation of the OR&A process requires a rigorous monitoring system which is both easy understand and intuitive to use. If the system does not have these attributes, it is unlikely to be used to its full potential and may not then achieve its core purpose.

The OR&A system referred to in this book, a product of Operations Readiness & Assurance (OR&A) Ltd., is a combination of technologies and methods which, although not new in their own right, are used in such a way as to present the subject matter in such a way that it is easy to follow for someone who is relatively new to the discipline, yet complex enough that experienced users still value the information it provides and are therefore understand the need to retain the systematic approach to ensure consistency.

The monitoring system is a stand-alone desktop software application that allows the user to record progress, comments, potential lessons to be learned, peer advice and any other pertinent information related to an Activity or Task.

The application can then generate detailed reports combining any of the information from the structured OR&A system and the user entered data, combined with the chronological metadata automatically attached to each piece of information.

What this means, in plain English, is that weekly and monthly reports, progress reports, status reports for task, activity and deliverables etc. can be automated, saving a great deal of time and effort.

The standard output from the application is used to present a clear overview of the implementation status of OR&A on a project. It can be viewed in a number of levels of detail making it an effective monitoring tool for any level of management or analysis.

In my experience (more than 12 years) of building and implementing OR&A systems for Royal Dutch Shell and other international Oil & Gas companies, I have yet to see another methodology for a monitoring system which can achieve all of these things in such a user friendly yet comprehensive manner.

8.2 OR&A System Outputs

Producing output similar to that generated instantly by the OR&A Application is obviously possible, but resource intensive and therefore not cost effective (or even practical).

The application can generate a full overview of OR&A status as a spreadsheet (if required) and can also generate a range of pre-defined reports in a matter of seconds with a single mouse click.

However, because the application is based on an SQL database, there are a number of other outputs which are of immense value to the OR&A Team in implementing the process.

8.2.1 Preparing an OR&A Plan

Typically, the OR&A Plan for a project would be prepared by the Project Planner based on information provided by the OR&A Manager/Lead for the project. This would involve several iterations by both parties and would rely on their experience and expediency to achieve a comprehensive result. The duration of this process could be measured in man/weeks.

Using the OR&A Application a comprehensive OR&A Plan, in the form of a Microsoft Project file, can be generated in a few minutes for the entire project. If the project uses Primavera as its preferred planning tool, the Microsoft Project file can be easily and quickly imported into Primavera.

The OR&A Plan generated from the OR&A Application retains the entire Work Breakdown Structure from the OR&A System down to task level, numbered accordingly. The Project Planner need only align this with the project dates and milestones to produce a comprehensive and detailed OR&A plan for the project, thereby incorporating a vast amount of detail into the Integrated Project Plan.

The only remaining activity is for the OR&A Manager/Lead to adjust the resources required for each activity (to address the magnitude and complexity of the Project).

From this information, the Project Planner can produce a detailed cost estimate for OR&A manpower and accurately identify the resources required to complete the work.

The OR&A Manager/Lead can then add the material resources and other costs to this very accurate estimate to produce an accurate cost estimate for implementing the entire OR&A Process on the Project.

This estimate can be quickly and easily revised in the same manner.

8.2.2 OR&A Spreadsheet

For those who only feel comfortable with a detailed spreadsheet, this can be easily generated from the OR&A Application in a few minutes. The spreadsheet produced allows the detail to be expanded or shrunk to display the required level of detail, from Group down to Tasks.

The current Traffic Light information, including the % completion values and any comments recorded are also included on the spreadsheet.

8.2.3 The OR&A Dashboard

The OR&A Dashboard is the 'shop window' of the system and allows the user to browse quickly through the entire OR&A hierarchy. It is so user friendly that users do not need a great deal of experience to use the system and can find a specific Task in the entire system in just a few clicks.

Basic progress is recorded using radio buttons, tick boxes and drop down selections where possible to minimise the effort/time required. It only takes perhaps 30 minutes each day to record the progress of an entire OR&A Team on a project and a matter of moments longer to generate the reports.

The screen displays are intuitive and users can be entering detailed information very quickly.

Another feature of the system is that it is small enough to be sent by e-mail, allowing the recipient to generate his own reports in a remote location, such as Head Office.

8.2.4 Tailored Reports

The system is supplied with a number of pre-defined reports which will be sufficient for most users; however, tailored reports can be created from any of the information held in the system.

Generic reports (supplied) include:

- Weekly/Monthly Reports
- Weekly Monthly Change Reports (for Planner)
- Deviations Report
- PEER Assist Report
- Potential Lessons Learned Report
- OR&A Spreadsheet
- Microsoft Project Plan

8.2.5 Providing Assurance

Any level of assurance, as explained earlier in this book, requires a robust approach and the data to be structured and measurable such that it can qualify that stated value of assurance. It also requires the outcome to be communicated in a form that can be easily understood.

The target for any assurance system must address the following parameters:

- The target value (i.e. what constitutes 'Readiness for Operations'),
- A mechanism to measure the data and report the outcome in a meaningful way.

Assurance, in the context of OR&A, is about providing assurance to the stakeholders (and other interested parties) that the project will culminate with the asset in a state of Readiness for Operations, as defined earlier in this book.

For that assurance to be 'meaningful' it must be measured against the project timeline, providing that assurance at any point in the project.

The output from the tool must indicate that, in respect of OR&A, the project is where it should be in relation to the Project timeline and the Integrated Project Plan.

Typical Assurance

Typically, that assurance is based on some form of representative sample, generally obtained by some form of review or audit, subsequently reviewed by a Subject Matter Expert (SME) who then provides an opinion of the current level of assurance.

Often, because of other constraints on time and project/review resources it is not possible to examine every aspect of the project. This reduces the size of the representative sample and introduces variations in the quality of the data in the form of opinions (some of which may have a vested interest in a certain outcome) which undermine the validity of the value of assurance produced.

Consequently, the validity of that value of assurance is only as good as the sample data and the quality and competence of the participants involved.

Real Assurance

A system such as that provided by the OR&A Application offers the only satisfactory solution to this issue, providing the necessary robust structure, measurement mechanism and validated output in an easily communicated way.

- First, for 'Real' assurance to be provided the sample data must be expanded to include all of the data available.
- Second, this data must then be rigorously evaluated to produce the output (i.e. the stated level of assurance).
- Third, the outcome must be communicated in a way that is meaningful yet easily understood.

As explained previously, it is always possible to prepare detailed reports manually, though this is resource intensive and introduces the possibility of errors and omissions, however, the OR&A Application can generate detailed, pre-formatted reports instantly, removing the possibility of errors and omissions.

The 'real time' output from the OR&A Application displays the status of the implementation of OR&A in the Project on a single dashboard screen.

The level of assurance is obtained by being aware that the application is displaying the actual status of every monitored item against where it should be on the Integrated Project Plan in the form of a 'Traffic Light' indicator.

Because it is possible to move down through successive levels of detail within the system using just a few mouse clicks, the actual status of any Discipline, Activity or Task can be examined and the actual completion of each displayed in detail.

Further Advantages
Because the fully populated OR&A Application can be sent by e-mail, the reviewers can conduct a first-pass (desktop) review of the content from the main office location using as many Subject Matter Experts as they feel is necessary.

When this review is complete and an interim opinion has been reached, (based on the actual status of OR&A in the project as reported by the content of the OR&A Application), the actual review team can be tailored and the most appropriate personnel despatched to conduct the actual review.

This has the advantage of being able to reduce the time taken for the on-site review and the inconvenience to the project, by focussing the attention and skills of the review team on the pertinent subjects.

To an experienced OR&A Practitioner, the various outputs of the application provide a great deal more than a basic spreadsheet could ever record and many are able to analyse the content to identify more complex issues that could be expensive to resolve if ignored until asset handover occurs.

9 Lessons Learned & Best Practices

9.1 Lessons Learned

9.1.1 Definition of a Lesson Learned

There appears to be an on-going discussion in the industry regarding the correct title of this subject. Should it be 'Lessons Learnt' or 'Lessons Learned'?

In describing lessons which have been identified during project activities that have then been re-introduced into a subsequent project activity, so as to prevent a repeat of a previous error, or whereby a subsequent project learns from those lessons, it is most definitely 'Lessons Learned'.

The Lesson(s) are the subject of the title and the word 'Learned' is attached as an adjective to describe those Lessons, (i.e. lessons that have been learned by experience or study).

Example: Here are some lessons learned from the recent project, which are particularly relevant for anyone participating in a similar project.

9.1.2 Identification of Lessons Learned

Identification of a potential lesson is usually achieved by one of the following means:

- The consequences of the error, oversight or omission are manifested in the outcome of the project (failure),
- The consequences of the error, oversight or omission cause loss or harm (incident),
- The error, oversight or omission is identified by a project activity such as a health check, review or audit.
- As a result of a good idea or an innovative concept that subsequently resulted in a positive outcome for the project.
- Formally, in the context of a Lessons Learned Workshop or similar activity.
- Informally, as part of a discussion that is not part of a formal Lessons Learned exercise (see Knowledge Management).

Figure 15 - Lessons Learned & Best Practices

9.1.3 Potential Lessons

An error, oversight or omission in a project or project activity, that is subsequently identified or discovered by some means, cannot simply be considered as a 'Lesson Learned' because, at this point, nothing has yet been learned. Until it has been developed, it can only be truly called a 'potential' Lesson.

A large number of potential Lessons may be identified on any given project. When these are developed, detailed, analysed, categorised and evaluated, a number of Lessons Learned will be created.

Lessons Learned produced from project should be included in an administration, storage and retrieval system. Not all Lessons Learned will be applicable to every project, so some method of context sensitive sorting and evaluation methodology or system is also necessary.

The difficulty lies in categorising and quantifying that 'potential' lesson and arriving at some form of mitigating action which could be applied to prevent its occurrence. To become a true 'Lesson Learned' therefore, the point at which that mitigating action needs to be implemented must also be identified.

9.1.4 Classification

Determining what mitigating action can be implemented may be difficult enough, but to turn this into a Lesson Learned which can be generically recommended for other similar projects could be equally difficult.

The system used by OR&A Ltd. allows the user to identify potential Lessons Learned at both Activity and Task level, providing a great deal more context than would normally be provided.

Because of the way the OR&A System is robustly structured, when a potential Lesson is registered against an Activity or Task, this indicates the context, the applicable discipline or process area involved and provides a time line reference against the project timeline by default.

The additional detail from the Task in hand at the time and the circumstances which led to the identification of the potential Lesson allow the classification of the potential Lesson to be more easily determined.

The project timeline reference provides a reference-point when considering the implementation of any mitigating action and the magnitude of the error, oversight or omission at that point will go some way to identifying the cost and required effect of that mitigating action.

It should also be remembered that Lessons Learned are not only identified from errors, oversight or omissions, they are often the outcome of an activity that was very successful. A potential lesson could be the result of a good idea or an innovative concept that resulted in a positive outcome for the project.

9.1.5 Collection/Retention/Storage

Ideally, collection of a potential lessons needs to be done as soon as possible after it is identified to ensure it is recorded as accurately as possible. These potential lessons should be stored in a common location which allows them to be easily retrieved at a later date.

On most projects, potential lessons are also collected in a formal Lessons Learned Workshop, usually at the end of each phase and at the end of the project during the close-out process.

As previously mentioned, the system used by OR&A Ltd. allows the user to identify potential Lessons during the process of using the OR&A System and these can be retrieved, as a report, at any time.

9.1.6 Retrieval/Application

When a company has managed to assemble a collection of Lessons Learned, the next problem is the searching the database of those lessons and the retrieval of those that are appropriate or applicable.

Without accurate and context sensitive evaluation of those lessons, actually deciding which ones to apply in any given project can be both time consuming and costly.

Needless to say, OR&A Ltd. has a system which allows context sensitive search and retrieval of Lessons Learned using a multi-term searchable database to identify applicable lessons.

Within the OR&A Ltd. OR&A System itself, the facility exists to add references to specific Lessons Learned to particular Activities and Tasks in the OR&A System, thereby automatically promoting applicable lessons at the appropriate juncture.

The system also includes additional functionality that allows any Activity or Task to be tagged with PEER advice. The PEER advice could be an advisory note, a warning, a recommendation or a source of further information. These can be easily added by any user.

9.1.7 Promotion of Lessons Learned

Occasionally, a project will identify a Lesson Learned which is also considered to be a Best Practice, (i.e. something that should be included in every project). These will be clearly documented and the processes and procedures of the company should be amended to include it.

9.2 Best Practices

9.2.1 Identification

Identification of a Best Practice is usually achieved by one of the following means:

- From Industry in general, in the form of ideas or concepts suggested in trade publications, by regulators or industry experts.
- As a result of a good idea or an innovative concept that subsequently resulted in a positive outcome for a project.
- Formally, in the context of a Lesson Learned that is considered to be so good it should be adopted as a Best Practice.

9.2.2 Retrieval/Application

Best Practices should be incorporated in the processes, instructions and procedures of the Company to ensure it is shared between Business Units (Projects) carrying out similar activities.

It would also be pertinent to manage these in the same way as Lessons Learned (since they are effectively an extension of the same concept).

9.2.3 Promotion of Best Practices

Originally, most Industry Standards were developed from Industry Best Practices, methods and systems, because they were generally considered to be the safest, most efficient and profitable way of doing business.

Some of the Best Practices adopted by the Company could also fall into this category and therefore be adopted as Company Standards.

A Company may also develop a new process, system or methodology that becomes a Standard.

9.3 Skeleton Documents

9.3.1 What is a Skeleton Document

Essentially, a Skeleton Document is what the name implies, the 'bones' of a document that requires the 'flesh' to be added.

It is common for a Company to use a Document Template to produce a standardised style and presentation for a core business document. The 'Skeleton' concept takes that a step further by retaining the headings and structure of the document to ensure completeness and a standardised presentation.

This concept is further enhanced by including guidance text under each key heading that explains what should be included in each section of the document and an indication of the depth and detail required.

This not only enhances the quality, completeness and standard of the document in question, but increases the efficiency of the document production process by reducing the amount of time wasted in 'prettification' or formatting and tinkering with the document layout.

9.3.2 Creation

Skeleton documents are most easily created from Best Practice examples, or alternatively by a Subject Matter Expert or Technical Authority who is capable of setting the standard required for a given document.

9.3.3 Retrieval/Application

The Skeleton documents should be stored in an electronic document library as with any key project processes, procedures, instructions and documents.

Reference can then be made to the appropriate Skeleton documents in the same way as other business documents are referenced from the OR&A System.

9.4 Knowledge Management

'Knowledge and wood should not be much used until they are seasoned'

Oliver Wendell Holmes, 1809 -1894

9.4.1 Explicit Knowledge

The knowledge acquisition process understandably begins with the explicit or tangible knowledge in the capture of standard corporate documentation such as strategies, plans, standards, processes and procedures (most of which are usually recorded somewhere anyway and often used frequently in the normal day-to-day execution of business activities).

The assumption made by most companies or organisations is that they have all the knowledge they need in some form or other and that it just requires 'sorting out'. That 'sorting out' process can be a task of gargantuan proportions and so must be carried out in a controlled and structured manner.

Thy must first have a structure in place to ensure the 'sorted' data is stored in a consistent manner that ensures the context of the information is retained. However, the point is that the traditional data collection process is only valid for the explicit knowledge of the company.

9.4.2 Tacit Knowledge

More difficult to acquire is the tacit or intangible knowledge of the company or organisation since this knowledge takes the form of things we know, but are not consciously aware that we know.

These may be life experiences, ideas, insight or intuition and also include general know-how or other transferable skills and techniques.

This information is in the minds and memories of the employees of the company or organisation and can only be harvested by means such as discussions, workshops, interviews etc. (in other words by human interaction on a face-to-face basis).

This needs to be recorded and subsequently translated into explicit knowledge, proofed and validated by other suitably qualified and experienced people.

Other forms of communication that may facilitate accumulation of tacit knowledge/intangible information by providing an informal or semi-formal vehicle for communication include various forums such as facebook, twitter, blogs, chat-rooms or informal web communities such as Linked-In.

9.4.3 Implicit Knowledge

Yet more difficult to acquire or capture is implicit knowledge. By the very nature of it, this type of information has probably not even been considered seriously or thought about consciously by the holder, which makes it difficult to even describe, yet alone capture.

Often, such knowledge can only be collected in less controlled circumstances and this is where the novel concept of Communities of Practice (CoP), (an idea by Etienne Wenger and Jean Laval), has become popular.

The concept suggests that a deliberately contrived socially oriented gathering of like-minded individuals, (i.e. people with one or more areas of common interest, such as business, football, golf, motoring or just social interaction), when meeting in a social setting, are apt to informally discuss their various interests (in a less formal way that they would be inclined to do in a formal business context).

In such a deliberately informal context, unexpected and alternative ideas, irritations, minor complaints, radical thoughts and concepts that would normally remain unsaid or perhaps summarily dismissed or ignored in a more formal setting may be articulated, discussed or argued about. It is those thoughts and ideas that may eventually prove to be of the highest value to the company or organisation.

Recently developed web technologies such as Linked-In, Facebook and Twitter can also be utilised to provide the infrastructure to support a community of practice and promote the necessary informality (and anonymity, if required).

However, to be effective a Community of Practice must be managed, usually by a co-ordinator who is also an active member of the CoP and must encourage, nurture and facilitate but never attempt to force an issue or dictate an agenda. Every Community of Practice will have a natural lifespan and will only exist so long as the members wish it to continue.

9.4.4 Knowledge Management in the OR&A System

The way the OR&A System handles Knowledge Management serves to address all of the concerns explained in the previous paragraphs, in the following ways:

- The right information is provided in the right place at the right time to the right person due to the structured approach of the OR&A System to performing Activities and Tasks. This context sensitive approach is recognised as a very efficient primary method of Knowledge Dissemination.

- Further supporting information is easily accessible/retrievable by the user with the minimum of effort using the context specific hyperlinks from within the OR&A System.

- Best Practice examples demonstrate the standard, quality and completeness required for written deliverables.

- Skeleton Documents for key deliverables improve efficiency by speeding up the production of documents whilst standardising quality, presentation and content.

- Collection of potential Lessons Learned is facilitated at any point during the Project, automatically categorising and time-stamping the point in the Project at which the effects of the issue leading to the identification became apparent.

- The staging of Lessons Learned workshops to gather potential Lessons Learned at key points in the Project is a key component of collecting knowledge and creating Knowledge Base.

- The web-based approach of the OR&A System allows Lessons Learned to be embedded in future projects by editing of the content from a central location which is immediately accessible to all users.

- If the OR&A Lessons Learned System is used, Lessons Learned can be easily added, stored, managed, searched, sorted and retrieved from the Lessons Learned Library.

10 Further Information / Assistance

The author of this document and OR&A Ltd. are able to provide a great deal more in the way of services, support and guidance for Companies who would like to create or develop an OR&A capability of their own.

This includes the following services:

- Provision of OR&A software and systems,
- Provision of Briefings and Training on OR&A,
- Tailoring of OR&A software and systems with User Specific tasks and activities,
- OR&A Reviews and Audits.

More information is available from the Company website at:

http://www.operationsreadinessandassurance.com

or

http://www.or-and-a.com